PERGAMON INTERNATIONAL LIBRARY
of Science, Technology, Engineering and Social Studies

*The 1000-volume original paperback library in aid of education,
industrial training and the enjoyment of leisure*

Publisher: Robert Maxwell, M.C.

THE DIMINISHING
RETURNS OF TECHNOLOGY

THE PERGAMON TEXTBOOK
INSPECTION COPY SERVICE

An inspection copy of any book published in the Pergamon International Library will
gladly be sent to academic staff without obligation for their consideration for course
adoption or recommendation. Copies may be retained for a period of 60 days from
receipt and returned if not suitable. When a particular title is adopted or recommended for
adoption for class use and the recommendation results in a sale of 12 or more copies, the
inspection copy may be retained with our compliments. The Publishers will be pleased to
receive suggestions for revised editions and new titles to be published in this important
International Library.

THE DIMINISHING RETURNS OF TECHNOLOGY

An Essay on the Crisis in Economic Growth

by

ORIO GIARINI

and

HENRI LOUBERGÉ

Translated by MAURICE CHAPMAN

PERGAMON PRESS

OXFORD · NEW YORK · TORONTO · SYDNEY · PARIS · FRANKFURT

U.K.	Pergamon Press Ltd., Headington Hill Hall, Oxford OX3 0BW, England
U.S.A.	Pergamon Press Inc., Maxwell House, Fairview Park, Elmsford, New York 10523, U.S.A.
CANADA	Pergamon of Canada Ltd., 75 The East Mall, Toronto, Ontario, Canada
AUSTRALIA	Pergamon Press (Aust.) Pty. Ltd., 19a Boundary Street, Rushcutters Bay, N.S.W. 2011, Australia
FRANCE	Pergamon Press SARL, 24 rue des Ecoles, 75240 Paris, Cedex 05, France
FEDERAL REPUBLIC OF GERMANY	Pergamon Press GmbH, 6242 Kronberg-Taunus, Pferdstrasse 1, Federal Republic of Germany

First edition 1978

British Library Cataloguing in Publication Data

Giarini, Orio
The diminishing returns of technology. —
(Pergamon international library).
1. Economic development   2. Ecology
1. Loubergé, Henri
301.31   HD82   78-40569
ISBN 0-08-023338-4 Hard cover
ISBN 0-08-023337-6 Flexicover

*Printed in Great Britain by William Clowes & Sons Limited
London, Beccles and Colchester*

"Le présent serait plein de tous les avenirs,
si le passé n'y projetait déja une histoire."
(The present would provide an infinite choice
of futures if it were not already the projection
of a story begun in the past.)

André Gide, *Les Nourritures Terrestres*

"If the subject-matter of a science passes
through different stages of development, the
laws which apply to one stage will seldom
apply without modifications to others."

Alfred Marshall, *Principles of Economics*

Contents

Preface:
The Biography of This Book

by Orio Giarini

This book is a brief statement of ideas and experience accumulated over a period of nineteen years in an effort to understand what industrial society is and where it is heading.

The inquiry started with direct experience and observation of industry. This was then supplemented by reading, discussion and teaching work.

1. The debate on the post-industrial society

The process was started by Daniel Bell's first European lectures in Salzburg in the summer of 1959 on his conception of the post-industrial age. I had a strong feeling at the time that he was trying to counter Marxism by depriving it of its sociological justification, i.e. by showing that the post-industrial world was tending to eliminate industrial labour and, with it, the lever for the switch to socialism. This was linked to the idea that progress and modernity meant the end of "ideologies".

It looked suspiciously like a conjuring-trick: a change of production structure does not necessarily abolish conflict and exploitation of man by man. Surely, human nature will always find a way of preventing redistribution of power and wealth, regardless of the sociological substructure. In all human history, no political structure has been able to avoid the gradual loss of its noblest and most egalitarian aims — though this is of course no reason for giving up the attempt.

So I felt that Daniel Bell's sociological analysis was strongly coloured by ideological motives and did not provide an adequate picture of a "production structure" actually supplying the things that we use and consume. His announcement of the "death of ideologies" seemed inconsistent with his motivation, but this might well be the result of a kind of American cultural egocentrism. I had become convinced during a spell at an American university that every society has in fact an ideology. There

1

may be cases where this is not explicit because it is shared by the whole society and there is no contrasting ideology to make it evident.

The history of the American Frontier shows that, while unoccupied or sparsely occupied lands were available, social conflict could always resolve itself through geographical dispersion. Hence, the ending of the frontier epoch marked the beginning of a new chapter in American history, making it increasingly necessary to spell out an ideology under the name of a "system of values". The two sides in a conflict now had to compromise, being no longer able to opt for unrestricted freedom by moving elsewhere.

From this angle, Daniel Bell's way of thinking struck me as being in reality the conservative answer to the same problems that faced John Kennedy when he tried to propose a new frontier ideal. On one side was the hope of maintaining the freedom from "ideological" constraint of the old frontier society. On the other side was an attempt to make up for the lack of a territorial frontier by giving Americans new frontiers for the imagination. Unfortunately, even the mental appeal of the new space frontier cannot replace the earlier geographical reality. In spite of what many sociologists and historians have affirmed, the United States seems likely to move nearer to Europe and the rest of the world in its formulation of an ideology or explicit system of beliefs, rather than the other way round. The no-ideology dogma is probably just as mistaken and destructive as all ideologies that have given up the constant search for a better balance of values and have become dogmas.

Daniel Bell's zeal in criticizing Marxism did not greatly bother me. The impact of this type of problem on my generation (and this will have one day to be accepted) has not been the same as it was on previous generations in Europe. Now that Marxism has played a part in the formation of European culture and is still a significant element in its stock of ideas, creativeness in the social and intellectual fields must inevitably go beyond the interpretation of texts.

My chief interest at the time was in obtaining a closer view of how the machinery of production and distribution actually operates. In Italy, my university had provided the broad outlines and concepts, especially in the area of production functions where capital and labour are ingeniously combined in equations; but I needed to learn how the economic process was carried on in practice.

2. The constraints of modern industry

After the excursion into the economic theory of production factors and Daniel Bell's sociology, what particularly attracted me were the details, and this coincided with the need to take a job. In 1959, post-war reconstruction in Europe was nearing completion and the giant industries ruled supreme; so I went into the chemical industry to promote some new products in which certain companies were placing high hopes.

The experience of meticulous and time-consuming work with production and quality control engineers gave me an idea of the problems and delays in getting a new product off the ground. The task of promoting a new product consisted in defining — in view of a particular use and bearing production cost in mind — problems to be solved by the engineers, e.g. showing that a type of plastic sheeting would have no smell and was impervious to oil, that another product would stand up to heat or light, or that a synthetic yarn would withstand twisting in textile machines. The striking thing was the number of failures, the time needed to develop a moulded cup using a new type of plastic, and the importance of operating know-how at each stage of manufacture.

At that time the chemical industry was still in its heyday but the period of major disappointments had begun. Every sizeable company was looking for a way of achieving a brilliant success by developing new products, in the hope of repeating what Du Pont de Nemours had done with nylon, I.C.I. with polyester, and Montecatini with nitric acid. The attempt was beginning, however, to prove very costly because of the number of setbacks in the lengthy process of development.

Apart from this, sudden rises in the market price of materials frequently occurred. On the international market for nitrogen fertilizers, I was surprised to see prices double, treble or go even higher, without any political change whatever being involved. Where was supply and demand equilibrium if this could occur? While the monopolistic factor might be operating in certain limited cases and in very unusual conditions, the sharp variations in markets such as fertilizers seemed much more likely to be due to structural factors — which are discussed in this book in connection with the growing vulnerability of the economic system, and which should not be underestimated in the light of more recent substantial increases in oil prices.

3. Some years in research and development

After some years in industry, my questions as to "how it works" had undergone a change of direction. Clearly, the source of the modern industrial process lay in science and technology, and it was necessary to understand how this link-up took place.

An opportunity to join the Battelle Institute arose, and I decided to leave my sales management job in industry. I had gained some experience but with the same limitation as most people, i.e. I did not know how far it was a one-sided experience and how far it was representative of the logic of events in industry generally.

Like everybody else, I was fascinated by the latest developments in technology. In fact, this led me to make a study (alongside my research for the chemical, textile and engineering industries) and publish a book on the conquest of space,* in which I presented a picture of technology which the present book partly rejects.

Nevertheless, that first book already made the point that the technology employed had been primarily the outcome of ten years of sustained efforts in the utilization of *available* knowledge. This showed that the magic words "Science and Technology" or "Research and Development" were not an entirely accurate description of how the fairy godmother works. I later came to agree with parts of Thorstein Veblen's analysis of scientists and intellectuals, i.e. that their credibility with the general public depends on their use of incomprehensible language (showing that unlike ordinary mortals they can speak to the gods) and on their reproducing phenomena that no one else could (so proving their possession of magical powers). That is no doubt why science and technology, which started in the revolt against superstition, have in the last decade or so become the object of a strange modern need for superstition.

To come back to the techno-economic research at the Battelle Institute, this was a matter of evaluating the economic possibilities of new machinery and new processes. The years around 1965 were the last of the technological boom. No sooner had a project been completed for an ethylene or ammonia plant producing 500 tons a day than planning began for a plant to produce 1000 tons. The same was happening in aviation (up to the Boeing 747), in computers, and in other major industrial sectors. From the economist's point of view, it was a question of taking advantage of

* *L'Europe et l'Espace,* Centre de Recherches Européennes, Lausanne, 1968.

economies of scale. To finance and administer larger units producing at faster rates, firms had to increase in size through the process described in all the manuals on economics. These rarely made more than a passing reference to the fact that, beyond a certain point, any further concentration may be uneconomic.

Innumerable studies were based on this approach, but there were increasing symptoms of malfunctioning in the growth mechanism, and there were some cases where increased scale even led to diseconomies. Over-large scale multiplied the costs and risks at the entry and exit of the production process, especially because of warehousing and distribution problems. Any chance of selling a research project for an ammonia plant producing 5000 tons a day had disappeared, but there was a demand for projects to improve product storage and transportation systems. Similarly after the Boeing 747 came on the market, there was increased research on ground organization to cope with the flow of goods and passengers. Techno-economic research often showed that the gains from specialization and concentration in a manufactured product were offset by maintenance problems.

An outstanding example comes from the heyday of research on "disposable" products. When economists realized that it would be easier to increase a market by halving the life of the product than by doubling the number of buyers, no suggestion seemed too fantastic. One was the throw-away watch, which has been achieved, and another was the motor car with a large number of replaceable parts. The idea of disposable goods preceded by a very short time the ecological movements working in exactly the opposite direction.

Thus, the first reaction in the textile industry was that all laundering of linen in homes, hotels and hospitals could be eliminated. However, it quickly became obvious that the quantity of disposable sheets, etc., needed to replace the washable versions would create insuperable problems of storage in the home — every house would become a warehouse, and the washing machine would have to be replaced by a "compacting" machine for waste. The cost of having less washing to do would have been considerable and, in the end, involve more work, without counting the enormous waste of raw material.

This extreme case suggests a different approach to the "value" of products from that adopted in traditional economic theory. What matters to the individual is the value of a product in use, not the product as such. The question of how long the life of products should be is probably a very promising starting-point for the debate on ecological and economic requirements.

This brings us to the question of the use of energy. About ten years ago, another very popular idea was that there should be a move to methods of production using larger amounts of energy. It was expected that power would become cheaper with the development of nuclear reactors, which would enable the kilowatt/hour price to be brought down to the level of Norwegian hydroelectric power. Accordingly, research for the intensive utilization of energy was seriously considered. With very cheap energy it might become economic to extract aluminium from the poorer deposits, or to produce phosphorus using a thermal process in place of a chemical process.

Certain research projects were planned on the assumption of cheaper energy, but it began to appear that the new sources of energy would not be available for a long time and that prices would certainly be higher than the first estimates. During the period 1970−2 the problem of shortages of energy and increasing cost began to be mentioned in research reports, and even in the technical press − in the field of aviation, for example.

During the same period there was also a steady flow of indications that it would be many years or even decades before the new methods of producing, transforming, storing and distributing energy could become economically significant. Examples ranged from magnetohydrodynamics to fuel cells and from the utilization of oil schists and super-conductors to nuclear fusion power-stations.

Since that time, the obstacles in terms of rising costs and longer lead-times have become not less but more serious whenever a project is closely examined.

4. The Club of Rome

The foregoing was part of the background to the decision of a group of leading figures in scientific research, large-scale industry and the universities to set up an informal association − known as the Club of Rome − in a joint effort to gain a better understanding of the increasing number of disquieting factors that appeared to invalidate our experience with research and economic development in the 1950s and 1960s.

The Club of Rome was founded in 1968, with Aurelio Peccei and Alexander King as its prime moving forces; and its Executive Committee held its meetings for the first three years at the Battelle Institute in Geneva. The year 1968 is significant since student disquiet at the way in which the social and economic system was developing was shared by some of the more sensitive figures in national life. It should be said here that there was not a grain of truth in the allegation that the Club of Rome was part of a "conspiracy", either by right-wing or left-wing circles, or that it was financed by someone with partisan aims. It was an entirely spontaneous group of individuals seeking to understand what was happening and to stimulate thinking and research along new channels.

The Club was highly successful with the first project that it sponsored: *The Limits to Growth*, edited by Dennis Meadows and based on pioneering work by Jay Forrester.

Having closely followed this project from its preparatory stage onwards, I was struck by the enormous interest that it aroused and, later on, by the violence of the attacks on it, mainly from economists. The arguments cannot be gone into here, but the whole controversy seemed very much like a modern version of the trial of Galileo. On a rough estimate, 80 per cent of the criticisms were based on things which neither the Club of Rome nor Forrester had ever said or done. Obviously, some ideological chord in our society had been touched in the discussion of a scientific problem. Growth had ceased to be simply an economic concept and had gradually been transformed into a basic "ideological" assumption for political and social equilibrium in industrialized societies — apparently without the intellectual and political leadership being aware of this ideological distortion. A phenomenon that had been significant in a particular historical context had become an obsession. There was striking unanimity between the criticism of some political leaders on the extreme Left and Right. Like many others, they angrily rejected the idea of any decline in the economic growth rate since their strategies assumed its increase. I was personally appalled at the idea of society becoming stuck in a creed based on an endless extrapolation from the unique experience of a quarter of a century, instead of treating this as a working hypothesis. Economic growth had ceased to be regarded as a means and had become an end in itself, thus eliminating any question of choice between future goals. This makes one wonder whether there is not a parallel between the opposition to the Meadows report and the fate of the succession of movements after 1968 whose forebodings fell on deaf ears, except in the case of certain ecological movements.

For some time after the Meadows report, the debate centred on the external limits to growth, and the oil crisis accelerated the growing awareness of the problem and its dimensions. It seemed to me, however, that the question of internal limits was also important, especially as the economists' main argument was that technology and the mechanism of relative prices would successfully push back the external limits. The wonders of modern science and technology were well known. What reason was there for not believing that they would do even better and — well before any catastrophe — find a way of producing all the energy and raw materials needed at a reasonable price?

These appeals for confidence suggested that the possibilities of science and technology were no longer a subject for controlled analysis but were now an object of belief, verging at times on superstition. I maintain that the worst enemies of science and technology today are not those embittered persons whose fears are aroused by every new discovery, but the people who subtly spread the idea that science and technology produce miracles.

The present book had its beginnings in that debate. What I had directly seen of economic realities had gradually crystallized. These findings were now to be discussed, tested and set within a general framework during six years of teaching at the Geneva University Institute of European Studies. Throughout that time, the help of Henri Loubergé has been invaluable, and many ideas in the book are the outcome of our discussions and the intellectual stimulus that these provided. In the course of time it became our main concern to clarify the part played by science and technology in the economic and social development of contemporary civilization. We began to see how greatly present-day economic thinking has been conditioned by the *a priori* assumption about science and technology. Because of this, Keynesians and Monetarists (especially the former) have their eyes still fixed on short- and medium-term demand, whereas what we are facing is a reversal of the long-term trend, chiefly affecting supply. All the same, we are anxious to emphasize the importance of economic phenomena in the modern world and of economics as a methodology and social science primarily seeking to throw light on the mechanisms of resource utilization and allocation. All things considered, the central aim of ecology can only be to suggest a better kind of economy, and this is what makes the period ahead of us so stimulating for thought and action.

5. Specialization and the cultural background

In the many lectures on these questions that I have given over the past five years, mainly to practitioners and educators in industry, I have often found that what I was saying was nothing really new for my audience. I was surprised to find that, in industries with which I was not familiar, the theory of decreasing returns of technology was operating to a much greater extent than I could have imagined. But I also noted that modern society seems perfectly able to function with economists and politicians believing in the power of technology, while simultaneously the experience of practitioners in their different spheres contradicts that belief.

One of the reasons is that there is little generalization from individual experience. Managers and engineers in one sector are unlikely to compare notes with those in another sector; each regards his particular experience as having no wider significance. Furthermore, where the general view is that technology is the way to success, any individual whose experience conflicts with this is reluctant to reveal that his situation and work do not match the standard of reference. A generally accepted idea can thus promote a sort of social alienation where the individual keeps his experience to himself, with all the loss that this involves in the area of "job satisfaction". Noting how incompetently developments in technology are reported, the industrialist or engineer usually shrugs it off with a smile and decides to stay within his own specialized field. One may respect this reaction, but it leads to under-estimation of the extent to which "imaginative" reporting is helping to create an ideological reality. The general reader is told that "aquaculture" will soon be able to increase fish production considerably, so he thinks that the food problem is at last being solved by technology. He forgets that, since Roman times, aquaculture has been "discovered" at increasingly closer intervals, and that this was already being presented as a "new" development five years before. Anyone can check that we are being bombarded with this type of news if he re-reads an item from a few months or years before. This all goes to show that the gap between specialist knowledge and the general culture is widening, and we believe that it partly explains the feeling of crisis and dissatisfaction in modern society.

Seen in relation to what has been said in this "biography", the final book is quite brief and probably much shorter than it should have been. While the ideas put forward in it are the result of much reflection and discussion, the text was written by the authors as and when time could be spared from their main occupations. It is hoped that the book will stimulate fresh discussion on the vital problem of adapting the economy to the changing character of civilization and to the problem of the external and internal limits to economic growth. We would like to assure the reader that the form of active pessimism involved in pointing out the limits is in effect equivalent to optimism. Economics has an essential role to play, and one that is much more significant than might be supposed from the current disillusionment at the inability of economic models to forecast correctly in recent years. The approach in the book is above all one of economic analysis, though a final chapter was needed to extend the analysis to the political aspects. And this brings us to the final stage of this "biography".

6. The federalist option

For a number of years I was in charge of the General Secretariat of the European Federalist Movement. I have often wondered since why the movement never reached the "take-off" stage, especially after the great crisis in 1954. What the federalists proposed had many attractions, but they were never able to break through on the political level in the way that the Club of Rome did on the cultural level — or even the ecologists in certain national elections.

European federalism offers a way of overcoming national exclusiveness and at the same time a way of safeguarding European independence and its distinctive national and cultural features. In the contemporary situation, it at least corresponds to a necessity founded on common sense. Those who argue that independence is solely a matter for the individual nation-states are in fact helping to make Europe and their own countries increasingly dependent. On this point, the federalists have the right approach, if Europeans genuinely want to construct the Europe of the future on a solid basis of realities.

However, the problem of European unity is not simply a question of institutions and foreign policy. It is also bound up with the over-all evolution of European society. Only a small number of federalists realize how these various aspects of the problem are interrelated. Nevertheless, when federalism becomes a blueprint for the wider society, it will have to take account of the ideologies and socio-economic realities that have dominated the present era. The essential factors that have operated over the last 200 years to make Europe what it now is were the progress of the industrial revolution and the spread of liberalism and socialism in the industrialized and industrializing societies. In the light of these developments, the ideas of federalists have often seemed to emanate from a pre-industrial society — and so, easily attributed to a hankering for the past — or else to reflect the feelings of certain revolutionaries (Proudhon, young Marx and others) on the question of local autonomy, or the doctrines of Christian pluralist sociology. Hence, while federalism has been a subsidiary element in projects for a European society, it has never gained acceptance as the basis or fundamental blueprint for a society in the context of the socio-economic problems existing at the time.

Since the industrial revolution, there has been a general movement in the direction of concentration of power. The "federalist" stages of some societies and nations have generally been stages on the way towards centralization. An increase in centralization is still today presented as an advance, even in Switzerland and the United States.

The fact that — as some political scientists point out — many nations, including the USSR, have a federal structure is beside the point. That is merely a nominalist classification. On the other hand, since we are certainly passing through a major structural change and have to use the term "post-industrial society" (without necessarily agreeing with Daniel Bell's use of it) to describe the socio-economic structure in which we are living, what lies ahead may not be a world that is more or less liberal or socialist or more or less advanced from the point of view of industrialization, but "something different".

If the "something different" is a society in which the production structure and search for freedom are moving towards a greater degree of pluralism; if "small" is not only "beautiful" but more efficient in relation to essential human needs; if world interdependence develops into a well-balanced* organic structure — then the federalist option may be the political answer for Europe in the post-industrial age.

This kind of federalism was born in 1968 with the Club of Rome and the ecological movements. It brings a message of hope and progress, in contrast to the lack of vision that has been partly responsible for the desperate nihilism of the frustrated. It is from this angle that the last chapter of the book puts forward a political analysis by way of extension of the analysis and hypotheses in the earlier chapters.

Geneva, March 1978

---

* Where there is interdependence between a stronger and a weaker partner, the second part of the word is what mainly applies to the latter.

CHAPTER 1

The Age of Uncertainty

The uncertainty of conditions in the future is one of the basic facts of the universe of space-time. When Man acquired consciousness of Change, he was brought face to face with this basic reality as a hazard of the human condition and a limitation on his freedom.

The fact that all human action takes place in an uncertain world places the individual in a situation of risk in his daily life. Specialists in the theory of risk and uncertainty, such as psychologists, economists and statisticians, generally make a distinction between "risk", where all the contingencies can be foreseen, and "uncertainty", where no such enumeration is possible.

Thus, uncertainty is essentially the realm of the unknown and unimaginable. The successor to Man in the evolutionary process on this planet, the forms of life in other stellar systems, the survival of the soul – these are archetypes of uncertainty. On a more matter-of-fact level, the state of economic and social organization a hundred years from now also comes under the heading of "uncertainty".

Every religion is above all a way of "managing" uncertainty – of reducing a multitude of potential futures for humanity and of individual destinies to a linear progression. It provides a meaning for life in this world, an answer to hopes for a hereafter, and an explanation for all that is unaccountable. Church membership is in fact one of the indicators used to measure risk aversion in a population (Greene [6])*.

This is why the renaissance and sytematization of the scientific attitude which took place in the West in the Modern period were an outstanding example of a revolution in the original sense of the word, i.e. a complete reversal of direction. The scientific outlook had never in Classical Antiquity or in the Middle Ages been so popular with a significant sector of the

---

* The figures in square brackets relate to the lists of references at the end of chapters.

population. The swing away from religion towards rational explanation of phenomena was equivalent to a refusal to "manage" uncertainty, coupled with a decision to make a direct attack on what is uncertain or unknown — these things being the starting-point or "raw material" of the scientist, who judges a theory by the number of doubtful points that it clears up.

With scientific progress, mankind has come to accept the lack of certainty in its environment and history as a fact — for example, Heisenberg's principle of indeterminacy [7]. For biologists such as Monod [12] and economists such as Georgescu-Roegen [4], chance is the ultimate factor in the nature of phenomena. All this has made it necessary to look for less ambitious ways of handling uncertainty.

In the field previously occupied by religion in matters of uncertainty, ideologies were able to develop in the nineteenth and early twentieth centuries. These did not offer any overall explanation of the universe, but "scientific truth" about the evolution of society and politics, and — in the case of Communism — a goal for mankind. The practical application of these ideologies was held back by other developments already in process, and by their own intrinsic limitations: in the social field also, uncertainty plays a bigger role than historical determinism was ready to allow.

In its withdrawal from management of the wider uncertainties, the Western world fell back on management of specific human risks, abandoning the earlier world views and so laying itself open to the consequences of a lack of goals to which Denis de Rougemont has drawn attention [13]. Risks affecting standards of living have become a major area of intellectual concern*, and management of these risks through economic policy has given rise to a mini-ideology, namely economic growth. Even the Communist movement, in its European version, has finally adopted mass consumption as the stock argument.

Economists have provided strong backing for economic growth, both in research and in practical matters showing once again that economics very

---

* It may be objected that economic policy considerations have always been important in the past. But this is only an illusion produced by our way of reading history. We tend to give most attention to passages of interest in relation to our current preoccupations, though these are generally different from the preoccupations of our ancestors. It is surprising to see, when re-reading the Classical Economists, how small a part of their work has been preserved and handed down. If they were here, they would no doubt be astonished to find that parts which they considered essential and developed very fully are now completely forgotten

naturally responds to the social preoccupations and ideas of each epoch, as can be seen from its history from Aristotle to the Mercantilists and from Adam Smith to Keynes. This has also meant that, when the ideology of economic growth was questioned, economics would be used as a weapon in the debate. As Serge-Christophe Kolm [8] has pointed out: "Economic science is neutral in the same way as a rifle — it is useful to the one who uses it."

The past ten years or so have seen the growth of an extensive literature drawing attention to the costs of economic growth, including some books by economists (Mishan [11], Scitovsky [14], Daly [2], etc.).

The debate on growth has led to the adoption of extreme positions on both sides, especially since the publication of the MIT report for the Club of Rome (Meadows [11]), with its timely reminder that this is a finite world. The backwash from this was inevitable, since even a minimal ideology cannot be expected to die quietly. All the same, it is a pity that a few economists abandoned analysis for wholesale condemnation on this occasion (Beckerman [1]).

The Meadows report concluded that — if there was no major change in the meantime — economic growth would cease in the middle of the next century owing to physical factors, i.e. depletion of natural resources and severe pollution of the environment. It indicated the emergence of what Jean Denizet [3] has called "absolute scarcities", in contrast to the relative scarcities dealt with in economics. For those who now believe the depression to be structural in character, the question is whether it has been caused by the sudden realization of the long-term limits to growth, or by other factors such as saturation of demand for consumer durables, the fall in the birthrate, the power of the trade unions, the downward trend in the rate of profit, political uncertainties, super-normal profits by monopolies, etc. Although there is a wealth of information, it is quite impossible to disentangle and weigh these different factors.

It seems to us more sensible to go back to the sources of the economic growth in the West over the past 200 years — especially in the past thirty years. We again find that the evolution of technology, the industrial revolution and the link-up between science and technology were the driving force of past growth. Starting from this point, it is possible to relate most of the above-mentioned growth-retarding factors — and many others — to a single original cause, namely a decline in the efficiency of scientific and technological progress. Such a hypothesis completely upsets the basic assumption on which economic analysis has been built up over the past twenty years or more, i.e. that technological innovation is a natural phenomenon, restricted only by the difficulties of its transfer to developing countries. The relevance of this hypothesis is not limited to the economic field. It may also have a bearing on the whole crisis of confidence in Western civilization from the beginning of the century – from Oswald Spengler to Ivan Illich. This technical civilization is the product of a systematic application of the view of the world inherited from the seventeenth-century rationalists. Hence, the first symptoms of a decline in the returns of technology could easily produce a loss of nerve among Western thinkers.

If this hypothesis is valid, growth is today coming up against certain internal limits. The external limits also exist, but, on the one hand, their existence is partly attributable to diminishing returns of technology (pollution) and, on the other hand, they in turn are having a retarding effect on technical innovation (the fear of exhausting natural resources). One might add in passing that, in these circumstances, there is little point in discussing whether or not technology will always find a way of pushing back the external limits. What proved true in the time of Malthus is not necessarily true today, especially as there is no counterpart to the introduction of potato growing and to the vast areas then waiting to be opened up to farming.

Nevertheless, the limit to which we are referring is not the inescapable technical barrier to the progress of technology described by Georgescu-Roegen [5] as follows:

> "The favorite thesis of standard and Marxist economists alike, however, is that the power of technology is without limits. We will always be able not only to find a substitute for a resource which has become scarce, but also to increase the *productivity* of any kind of energy and material....The idea is that technology improves exponentially. The superficial justification is that one technological advance induces another. This is true, only it does not work cumulatively as in population growth....Even if technology continues to progress, it will not necessarily exceed any limit; an increasing sequence may have an upper limit. In the case of technology this limit is set by the theoretical coefficient of efficiency" (pp. 16–17).

Nicholas Georgescu-Roegen is here giving the reason why technical progress is necessarily limited in the very long term (Carnot's coefficient of efficiency), but he does not mention that technology may, before reaching that limit, go through a process or processes of declining efficiency that are biological rather than physical in character.

Such processes have occurred in the past, and have been admirably explained by David Landes [9]:

> "Technological advance is not a smooth, balanced process. Each innovation seems to have a life span of its own, comprising periods of tentative youth, vigorous maturity, and declining old age. As its technological possibilities are realized, its marginal yield diminishes and it gives way to newer, more advantageous techniques. By the same token, the divers branches of production that embody these techniques follow their own logistic curve of growth toward a kind of asymptote" (p. 3).

Is the process of growth, maturing and decline through which industrial innovations pass transposable to a higher plane, i.e. technical progress in the aggregate? Is the principal factor in past growth, i.e. scientific and technical research, subject to the law of decreasing returns that applies to the other factors of production? These are the questions which will be considered more specifically in Chapters 4 and 5 of this book. We shall then examine in a concluding chapter the implications of this analysis for economic and social policy and political and social organization.

But before considering the limits which modern technology is facing, it may be useful to remind ourselves of what economic growth has come to mean in the contemporary world and what were the underlying factors that produced it (Chapter 2); and then to seek out the reasons why economists for a long time excluded the evolution of technology from the scope of economic analysis (Chapter 3).

It will be noted that this book is essentially the story of a marriage and a divorce. The marriage was between science and technology in the nineteenth century, and it enabled welfare and economic growth to progress side by side. The divorce is recent history. It is due to a divergence between economic growth and increased welfare because of a conjunction of phenomena suggesting that there are diminishing returns of technology.

18   The Diminishing Returns of Technology

References

1. Wilfred Beckerman. "Economists, scientists and environmental catastrophe", *Oxford Economic Papers,* November 1972, pp. 327–344.
2. Herman Daly (Ed.). *Toward a Steady-State Economy.* Freeman & Co., San Francisco, 1973.
3. Jean Denizet. "L'Europe étranglée", *L'Expansion,* September 1977, pp. 220–231.
4. Nicholas Georgescu-Roegen. *The Entropy Law and the Economic Process.* Harvard University Press, Cambridge (Mass.), 1971.
5. Nicholas Georgescu-Roegen. *Energy and Economic Myths.* Pergamon Press, New York, 1976.
6. Mark Greene. *Risk Aversion, Insurance and the Future.* Indiana University Press, 1971.
7. Werner Heisenberg. *La Nature dans la Physique Contemporaine.* Gallimard, Paris, 1962.
8. Serge-Christophe Kolm. *La Transition Socialiste.* Cerf, Paris, 1977.
9. David Landes. *The Unbound Prometheus – Technical Change and Industrial Development in Western Europe from 1750 to the Present.* Cambridge University Press, 1969.
10. Dennis Meadows *et al. The Limits to Growth.* Potomac Ass., New York, 1972.
11. Ezra Mishan. *The Costs of Economic Growth.* Pelican Books, London, 1967.
12. Jacques Monod. *Le Hasard et la Necessité.* Seuil, Paris, 1970.
13. Denis de Rougemont. *L'Avenir est notre Affaire.* Stock, Paris, 1977.
14. Tibor Scitovsky. *The Joyless Economy.* Oxford University Press, 1976.

CHAPTER 2

Lessons from the History of Economic Growth

Simon Kuznets [11] has defined the economic growth of a nation as "a durable increase of population and *per capita* output". His reason for using this narrow definition is that "a durable stagnation — or a decline — in population has rarely in the past two centuries been accompanied by an increase in *per capita* output".

In everyday use, however, the term "economic growth" is generally associated with an increase in the commercial production of final goods and services, or Gross National Product (GNP), in the course of a given period; and the term will be used in this broader sense in the discussion that follows.

2.1. Alternating pessimism and optimism in views of economic growth

In-depth study of economic growth has only become a major concern of economists in the last few decades. A recent work on the theory of growth [8] reminds us, however, that the topic was not unknown in earlier ages, when philosophers tried to understand the way in which wealth is produced and distributed. There are striking differences between writers of different periods as to the benefits of economic growth.

Probably the earliest writings on the subject are those from Ancient Greece. A twentieth-century reader is startled to find that, whenever Plato and Aristotle refer to growth, they stress the dangers and never speak of benefits. The philosophers of old rejected the idea of growth because of their conception of the well-being of society. In their view, this depended on permanence in the institutions and organization of the City. Any economic growth might undermine this permanency; it could not therefore be equated with an increase in welfare. For the ideal City in which the number of citizens was fixed and their occupations were precisely regulated by the laws, it was regarded in the same light as civil disorder.

In the Middle Ages and the centuries that followed it, the writers whose names we still associate with the study of economic phenomena took a different view. Ibn Khaldun, the Arab philosopher, and later the seventeenth-century Mercantilists and eighteenth-century Physiocrats, believed that growth was both possible and desirable. They devoted much effort to discovering its causes, among other things by studying the connection between economic and population growth. Towards the end of the eighteenth century, Adam Smith was the first to bring out the relationship between saving and capital accumulation.

Oddly enough, the nineteenth century in which the Industrial Revolution reached maturity was deeply pessimistic about economic growth. Growth was regarded not as undesirable in itself, but as a passing phenomenon that would inevitably be brought to an end by the normal working of the economy. This pessimism was derived from the works of Ricardo and Malthus. Both of these widely known writers believed that the law of diminishing returns in agriculture (through the bringing into cultivation of land of ever-decreasing fertility) and the law of equalization of profit rates between different sectors of the economy would lead in the long term to a general fall in profit rates. The result would be a lessening of incentives to invest and consequently a decline in capital accumulation, which is the driving force in economic growth. Although this scenario indicated a stationary state as the likely outcome for the industrialized nations, Ricardo's attitude was not one of resignation. He held that appropriate economic policy measures — above all, liberalization of trade — could help to stave off what he regarded as a calamity. He also believed that technical progress could push back the limits of the capital accumulation incentive, so enabling the industrial economies to delay the arrival of the spectre of a stationary state until the distant future.

The concept of a stationary state, i.e. a state in which production and population in a nation remain steady over time, is especially associated with John Stuart Mill [14]. Mill took over the Ricardian and Malthusian analysis but refused to regard a stationary state as a disaster. In a famous short chapter in the "Principles", he raised the question of the goal of economic growth, and his attack on the costs that growth involves might well have been written today.*

The thesis of a declining rate of profit was later taken over by Karl Marx, who assigned a major role to it in his forecast of the collapse of the capitalist system after being undermined by increasingly severe crises of over-production. Unlike Ricardo, Marx held that technical progress cannot significantly delay the process of deterioration since its aim is to economize labour. By doing this, it tends to narrow the base from which surplus value is derived, even though it increases the surplus value obtained per worker. Hence, the system is self-defeating.

In view of the progress made by the industrial nations in the second half of the nineteenth-century, one might have expected this succession of learned works throwing doubt on the possibility of durable economic growth to be followed by a new generation exalting the possibilities of growth. This did not happen. What is still remembered from the work of economists at the turn of the century mainly consists of demonstrations that an economy in which there is perfect competition inevitably moves towards an equilibrium point which is the optimum for the society.

Moreover, after the Great Depression in the 1930s there was a revival of "pessimistic" thinking. Economists began to take their cue from thinkers such as Oswald Spengler, Arnold Toynbee and Paul Valéry, whose study of the cycles in earlier civilizations provided the basis for their pessimistic philosophy of the Decline of the West. The "stagnationists" such as Alvin

---

* "It is only in the backward countries of the world that increased production is still an important object: in those most advanced, what is economically needed is better distribution, of which one indispensable means is a'stricter restraint on population...."

"If the earth must lose that great portion of its pleasantness which it owes to things that the unlimited increase of wealth and population would extirpate from it, for the mere purpose of enabling it to support a larger, but not a better or happier, population, I sincerely hope, for the sake of posterity, that they will be content to be stationary, long before necessity compels them to it...."

"It is scarcely necessary to remark that a stationary condition of capital and population implies no stationary state of human improvement. There would be as much scope as ever for all kinds of mental culture and moral and social progress; as much room for improving the Art of Living, and much more likelihood of its being improved, when minds ceased to be engrossed by the art of getting on. Even the industrial arts might be as earnestly cultivated, with this sole difference, that instead of serving no purpose but the increase of wealth, industrial improvements would produce their legitimate effect, that of abridging labour" ([14] Book IV, Chapter 6).

Hansen and Paul Sweezy, held at that time that consumption would level out in the long run so that the incentives to investment and economic growth would disappear.

Their reasoning was greatly influenced by J. M. Keynes, whose major work published in 1936 [10] had revolutionized the classical orthodoxies of economic policy. Keynes made the study of the consumption function one of the corner-stones of his analysis. In his model of economic activity, consumption increases with income but at a declining rate: when income rises, consumption and saving both grow but the proportion of income saved tends to increase at the expense of the proportion used for consumption. If this "basic psychological law" is transposed to the macro-economic level, then the share of overall consumption in the national income is bound to decrease as economic growth proceeds, with the consequences described by the "stagnationists".

Their mistake — for there was one — lay in combining two different things, i.e. household budget survey figures, which do indeed confirm the Keynesian thesis, and the trend in consumption over time. At any given moment, if one compares one income bracket with a higher one, then it is true that there is a lower average propensity to consume in the latter. On the other hand, if one period is compared with another, the proportion of income devoted to consumption in each social group is remarkably steady in the long term, even if the average income rises. This leads to the conclusion that the rate of consumption in each household above all depends on its relative position in the scale of incomes.* If this position does not change over time, the proportion spent on consumption is not bound to fall as income levels rise, and the argument of the "stagnationists" collapses.

All the same, the pessimism that still existed on the subject of economic growth just before the most extraordinary period of growth in human history is very striking. The opposition to the idea of continuous growth was just as vigorous as the campaign later on in 1972 to refute the arguments of those who dared to announce, in the middle of a boom, that there were limits to growth.

In actual fact, ideas changed very greatly in the space of two decades. Formerly regarded as an unstable and inevitably temporary phenomenon,

* See J. Duesenberry [3].

economic growth was seen after the Second World War as a universal remedy henceforth available to all nations of the Western world. Growth theory began to be one of the most influential fields of economic analysis. In the enthusiasm provoked by the amazing successes of the industrial nations, the overriding aim of economists was to discover the rules for balanced growth and avoid the Scylla and Charabdis of "stop and go".

Nevertheless, in the early 1970s after a decade devoted to the cult of growth, the MIT report [13] sponsored by the Club of Rome and certain other studies published in the United States [6], [4] struck a very different note by trying to show that growth could not be taken for granted and that humanity would one day have to relearn how to live without it.

As things now stand, there is no need to continue the controversy which for some years disturbed the small world of economists and politicians.* Moreover, the present section has already shown that the opinion of economists may vary at different times. It seems to us that the problem today is no longer one of being for or against growth. It appeared in this light when there was a prospect of a long period of growth before us. The question raised by the Club of Rome was: How long can growth continue? The basic question today is: Why is growth already running into difficulties? Hence, what we now need most is a better understanding of the way in which the process operates, in order to identify the areas in which economic policy can be most effective in given conditions. The emphasis is no longer on formulating the rules for optimum growth, but on analysis of the interaction of the factors in economic growth.

Since it is now accepted that technological progress has had a decisive influence on modern economic growth, the mechanism of this needs careful study so that we can estimate the impact that it may still have. Naturally, such an enterprise would be lacking in a solid foundation if it did not start out by taking due account of the history of the relationship between technology and industrial development.

2.2. The historical link between economic growth and technology

The date usually taken for the beginning of economic growth in Europe is 1750, which is also the accepted date for the beginning of the Industrial

---

* A special issue of *Futures* (February 1973) was devoted to the controversy on the "Limits to Growth" report.

Revolution in Britain. Thus, economic growth is closely associated with the growth of the industrial system.

It should not be thought, however, that economic development had been entirely absent in earlier centuries. Very little statistical information of an economic character is available for the Ancient World, but it is generally thought that certain periods of artistic and intellectual achievement, such as Imperial Rome and Greece under Alexander, correspond to periods of economic prosperity.

Moreover, an excellent outline by J. Gimpel [7] brings a reminder that Western Europe went through a period of intense technological activity between the eleventh and thirteenth centuries, which could claim to be the first industrial revolution. This period of expansion and technical improvements in the use of water power, in mining, in iron and steel metallurgy and in agriculture coincided, in the arts and philosophy, with the maturity of Gothic architecture as seen in the building of taller and taller cathedrals, and with the re-emergence of the critical spirit at intellectual centres such as Chartres, Paris and Oxford.

This exceptional period came to an end with the Church's condemnation of the new philosophy in 1277 and a series of natural disasters:

> "Within three years, from 1315 to 1317, the whole of Europe from Scotland to Italy and from the Pyrenees to the plains of Russia was struggling with appalling climatic conditions which led to the economic depression in the Late Middle Ages. This lasted for over 150 years until the Renaissance....The famines in the years 1315 to 1317 were so extreme that they terrified the European populations. They began in the summer of 1314 with torrential rains that flooded the crops in the North-Western flatlands. The price of corn and other foodstuffs rose steeply. Mounting inflation forced the king to bring in price controls for cattle and poultry. The bailiffs responsible for fixing and enforcing maximum prices were totally unsuccessful. The rise continued regardless of royal orders." ([7], p. 195).

The effects of the terrible famines due to natural causes were aggravated by the outbreak of the Hundred Years War in 1337, and of the Black Death which from 1347 to 1350 caused the death of (on average) 33 to 40 per cent of the population.

Such a period stands out in the history of the Middle Ages because of the length of the preceding period of prosperity and the severity of the depression that followed. But it must not cause us to overlook the underlying trend of a slow improvement in the standard of living. Between the year 1000 and the eighteenth century, income per head appears to have increased substantially, possibly threefold. But the trend was not apparent to contemporaries owing to the periodic famines which seemed to cancel out the progress made in the favourable periods. Economic factors were still viewed in the same way as in the Bible: Seven fat years followed by seven lean years.

The industrial revolution of the thirteenth century failed, for the level of development reached was not yet enough to set off a cumulative movement. Europe was vulnerable to extreme weather conditions because the general level of prosperity depended entirely on agriculture. After the depression in the late Middle Ages, the slow process of accumulation recommenced. In the eighteenth century Europe reached a certain degree of affluence and, as D. S. Landes [12] points out in his major work on technological change in the past 200 years, the conditions for the Industrial Revolution were now present:

> "Western Europe, in other words, was already rich before the
> Industrial Revolution — rich by comparison with other parts
> of the world of that day and with the pre-industrial world of
> today. This wealth was the product of centuries of slow
> accumulation, based in turn on investment, the appropriation
> of extra-European resources and labour, and substantial tech-
> nological progress, not only in the production of material
> goods, but in the organization and financing of their exchange
> and distribution" (p. 13).

This wealth meant that there was a demand for products waiting to be satisfied. Hence, the inventions of this period were put to industrial use instead of remaining, as in earlier centuries, mere sketches in dusty volumes.*

---

* The important part played by the revolution in agriculture, as forerunner to the Industrial Revolution, should be mentioned here. As Fohlen [5] points out, one cannot imagine the birth of an industry in a country where agriculture is at the purely subsistence stage.

We are often told that the Chinese invented the compass and gunpowder well before the Europeans, but did not know how to make the best use of them in terms of economic and military development. But Leonardo da Vinci was also remarkable for his many startlingly modern inventions, and these had no more impact on the contemporary economy than the Chinese discoveries had in the Far East. The inventions made in eighteenth-century England were not very different in kind from those known up to that time. In 1750, as in the Ancient World and the Middle Ages, technical progress was due to the intelligence of individuals aware of and interested in the world around them and the way in which things were moving. It so happened that one day their imagination suggested an idea and, having some technical knowledge, they were able to turn this into an invention.

The kind of technology which was put to use in the eighteenth century and which led to the Industrial Revolution can be exemplified by the picture of Denis Papin watching the effect of steam on the lid of a boiling kettle and getting the idea of the steam engine. No science is involved in this type of invention. To invent the steam engine, it was not necessary to know the structure of the water molecule but simply to be observant and imaginative. Similarly

> "Most of the innovations that transformed the textile industry in the course of the eighteenth century...were due more to "knack" than to any kind of research. The flying shuttle was invented in 1733 by a manufacturer of card strips for weaving machines, John Kay. The spinning jenny with which several threads could be produced simultaneously, unlike the single thread spinning wheel, was developed between 1764 and 1767 by James Hargreaves, a maker of weaving frames at Blackburn in Lancashire. The example most often quoted is Richard Arkwright, the inventor of the spinning frame, who was brought up at Preston to be a barber and wig maker....But the spinning frame only became operational with Crompton, a Bolton spinner and the inventor of the muslin wheel combining the advantages of the spinning jenny and Arkwright's spinning frame, which was later called the "mule" being a cross between the two."*

* Fohlen [5], p. 73.

Bertrand Russell [18] gave a good indication of the kind of machines used in the first industrial revolution when he pointed out that most of the machines in the strict sense involved nothing that merited the name of science. They were originally a simple means of making inanimate objects perform regular movemements which had previously been performed by human bodies, especially by human fingers. This was particularly true of weaving and spinning. Nor did science properly speaking play any great part in the invention of the railways and in steam navigation in its early days. In both cases, men made use of forces that were not in the least mysterious and, while their effects astonished them, there was nothing astonishing in the forces themselves.

The reason why the middle of the eighteenth century seems so fortunate from the point of view of inventions is that economic and social conditions at that time were particularly favourable for men of ingenuity. The various factors explaining why the "take-off"* occurred in Europe in the eighteenth century rather than at an earlier time in human history need no discussion here as they are fully dealt with in many works on the Industrial Revolution.** All writers agree on the importance of human and political factors in addition to the more specifically technical and economic factors. Some of the factors were peculiar to the eighteenth century. One is the emergence of a spirit of optimism which can be seen, for example, in the development of a philosophy based on the individual and Man's pre-eminence (Locke) within a framework of political liberalism (Montesquieu).

Others were deeply rooted in Western civilization, e.g. the Faustian striving for mastery of nature that emerged from Christianity, in contrast to the other major religions (except Mohammedanism) which sacralized the human environment. The approach to problems on the basis of reasoning and scientific method, inherited from the Ancient Greeks and restored to its place of honour by the Renaissance, also contributed to this evolution. A reference may also be made here to Max Weber's well-known argument that the typically Protestant virtues of austerity, thrift and hard work encouraged the emergence of an industrial middle class whose existence was a decisive factor in economic development in the seventeenth

---

* On the "take-off" and stages of economic growth, see the well-known book of Rostow [17].
** See, for example, Bairoch [1], Fohlen [5], and Landes [12].

and eighteenth centuries. However, there have been many criticisms of this thesis. An arguable alternative view is that the rise of the middle class promoted the advent of Protestantism in an environment which was increasingly rationalistic.

Nor must the favourable geographical and climatic factors be forgotten. Or the possibility that political factors may have operated as a "challenge" in Arnold Toynbee's sense — for example, the great invasions of the first millenium, and the unceasing wars fought in the Middle Ages and after, during the emergence of the nation states, by leaders seeking to consolidate these states with the help of powerful armies supported by prosperous economies.

Any consideration of the relative importance of the different factors would go outside the scope of this book. Let us leave the point to be settled by the economic historians, merely adding that the special reasons for the gap of several decades between the "take-off" in Britain and in the Continental countries is of great interest. On the whole it is accepted that, from the eighteenth century onwards, the British Isles benefited from the absence of barriers to internal trade and economic activity, and were not affected by the wars that disrupted economic activity on the Continent. Above all, the standard of living was higher and more equally distributed, and this relative affluence fostered psychological and social attitudes that encouraged the spirit of enterprise.

Generally speaking, therefore, there is a connection between the industrial revolution and the emergence of economic, human and political conditions in which *inventions of the traditional kind* resulted in *industrial innovations* leading to increases in productivity and therefore a fall in the relative prices of manufactured goods.

The repercussions of this radical change are well known. From the economic point of view, economic growth emerged as a tangible reality. Estimates of the national income in Britain indicate an average annual increase of 1—2 per cent in the eighteenth century and 2—3 per cent in the nineteenth century. Similar figures are generally given for France.

These figures may seem small compared with the records set in the period 1950—75, but they represent an extraordinary change of pace when judged by growth in earlier centuries. The effects on the population were immediate.* People lived longer, and infant mortality rapidly declined

---

* Indeed, some writers include population growth among the causes of the Industrial Revolution.

with the disappearance of famines and the combination of better public
health and a higher standard of living. The growth of the population was
accompanied by migratory movements set off by economic pressures, i.e.
the rural exodus in response to the manpower needs of incipient industry,
and emigration to the colonies in response to the home country's needs
for raw materials. All this took place against the background of change in
social structure from a society divided into nobles and peasants to a
pattern of capitalist and worker.

What is less generally known is that the Industrial Revolution also led
to an almost uninterrupted fall in the general level of prices, since the
technical innovations made for substantially higher productivity. Where
ten workers had been needed to mind 1000 spindles in a cotton mill in
1836, only three were needed by the end of the century. But the higher
productivity was hardly reflected at all in wage rates, since the mass of
workers were unorganized and formed a freely exploitable urban prole-
tariat. To some extent it helped to swell profits for reinvestment in the
business, which − as Bairoch [1] points out − was an important factor
at that time. But as competition was strong in a mainly atomistic market
and entrepreneurial freedom was a reality, improvements in productivity
were also partly reflected in falling prices. In the absence of wage increases or state intervention, there was little demand pressure on the
market, and so the fall could continue for a long time.*

The overall deflationary trend continued until it was interrupted from
1850 to 1873 by a rise in the money supply due to the arrival of large
quantities of gold from the United States. The amount of paper money
in circulation in France tripled between 1850 and 1870, and during this
period the growth of credit acted as a spur both to the development of
banking and to inflationary pressure. Whether this monetary incident
played any major part in the second wave of industrialization that followed
is still an open question. It is not strictly accurate to regard the nineteenth
century as a long period of regular economic growth due to the new technology and the advent of a powerful industrial middle class. Apart from
the short-term cycles analyzed, for example, in Gottfried Haberler's work
of synthesis [9], there were in the nineteenth century two cycles of the

---

* The economic philosophy of the period was largely based on Say's law, which
  is that − since supply creates its own demand − lasting overproduction cannot
  occur in an economy.

Kondratieff type, i.e. long-term movements in business activity. The first, starting in the middle of the eighteenth century with Britain's industrial development, is associated with the rise of metallurgy and, above all, the textile industry. This reached its peak in the first quarter of the nineteenth century and lasted until 1850, when a new cycle took over. This second cycle is generally associated with the arrival of the railways and the higher productivity and transformation of the economic structure brought about by this.* However, while not underestimating the influence of this innovation, it is important not to overlook the combination of factors that enabled it to produce its full effect. Once again, technological progress stimulated economic growth because the conditions were present for this to happen. With the increase in the money supply, demand revived and attracted technical innovation.** Simultaneously, the discovery of important coal deposits in the Rhineland and Westphalia and in the north of France enabled manufacturing industries to obtain inexpensive electric power. Furthermore, the spate of innovations induced by the industrial revolution had not yet reached its limit. The advances made in certain sectors that had benefited from the first innovations were reflected — at all the different stages in the transformation of raw materials — in further improvements to eliminate bottlenecks as these arose. The successes in the field of transportation in the middle of the nineteenth century (railways, steam shipping, canals) is best viewed in this light.

The figures quoted below give an idea of the intense development that occurred in the period before 1875.†

---

* Landes [12] (p. 196) points out among other things that the railroad provided possibilities for competition to operate effectively and so helped to squeeze out the inefficient regional monopolies that had been able to survive when transport costs were higher.

** The relationship between demand and technological progress has probably been an important factor in economic growth over the past 200 years. Fohlen [3] believes that "an innovation corresponds to an economic pressure or rather a "call" from consumers, and the research process interprets this need" (p. 83).

The question is whether this relationship is a stable one in the longer term, i.e. whether it is still valid today. If it were not, then the whole policy based since Keynes on stimulation of demand would be in jeopardy. However, the answer can only come from a deeper knowledge of the technological phenomenon.

† Figures quoted from Landes [12], p. 194.

	Railroad mileage (statute miles)		Coal production (1000 tonnes)		Raw cotton consumption (1000 tonnes)	
	1850	1873	1850	1873	1850	1873
Germany	3639	14,842	5100	36,392	17	117
France	1869	11,518	7225	24,702	59	55[a]
Britain	6621	16,082	37,500	112,604	266	565
Belgium	531	2335	3481	10,219	10	18

[a] 80,000 tonnes in 1872; 1873 was a bad year.

At the political level, the influence of Victorian Britain reached its peak in this period. The doctrine of liberalism was victorious in international economic relations, with a striking increase in trade between the European countries and between the home country and the colonies. In Germany and Italy, internal barriers were dismantled, and political unity closely followed economic union.

The period of diffusion and maturation of the Industrial Revolution had proved a brilliant success. The elation of contemporaries prevented them from seeing, however, that it also marked the ending of an epoch. In the industries which had taken the lead, "the giant steps were past. It was now a question of marginal gains in productivity, of filling in corners, of waiting for mechanical improvements to increase slightly the economic advantage of the new equipment....".* The foremost of these, the textile industry, had exhausted the possibilities created a hundred years before by the innovations of the Industrial Revolution.

The "primitive" technology had still something to offer since its geographical diffusion continued, and it still produced secondary effects in industry at all levels. This relaying effect took time before it had worked through. However, whilst this was taking place and social patterns and living standards were changing, there were the first signs of a technology designed on a larger scale and making use of the latest discoveries of science.

Although the transition to a new technology was probably not perceptible at the time, it had already begun and represented a concrete embodiment of the belief in progress that was current at the time. The advances that had been made in education had a considerable impact at this stage, as Landes points out:

* Landes [12], p. 211.

"At the middle of the century, technology was still essentially empirical and on-the-job training was, in most cases, the most effective method of communicating skills. But once science began to anticipate technique — it was already doing so to some extent in the 1850s — formal education became a major industrial resource and the continental countries saw what had once been compensation for a handicap turned into a signifi-cant differential asset" (p. 151).

The prime example of the transition is the technical advance that took place in the iron and steel industry at that time. Metallurgy is thus an exception to the rule that the successive waves of technical progress affect different sectors each time and promote them to leading industries. It had been one of the main instrumentalities of the first industrial revo-lution (though to a lesser extent than textiles) and it again passed through a lengthy period of technical upheaval in the second revolution. The con-struction, armaments and transportation industries were making heavy demands for steel, which could not be met with the puddling technique inherited from the eighteenth century. Hence, the incentive to develop a speedier process of steel production and so reduce the cost was parti-cularly urgent. This led to major improvements with the Bessemer con-verter (1856) and the Siemens-Martin furnace (1864). These were of much the same type as the improvements that had become familiar since the first industrial revolution.

Bessemer's technique was to blow air into the furnace in order to speed up the process of reducing the carbon content. Martin's idea was to use scrap iron to achieve the same result. The disadvantage of both pro-cesses was they did not eliminate the phosphorus contained in iron ore. As deposits of iron ore with a very low phosphorus content existed in very few places in the world (United States, Cumberland in England and the Bilbao area in Spain), the prospects for expansion in iron and steel were far from promising.

The solution was found in 1878 through the ingenuity of a police court clerk, Sidney Thomas, helped by his cousin Percy Gilchrist, a chemist in a Welsh ironworks. It consisted in putting basic limestone into the furnace of molten iron; this absorbed the phosphorous oxides produced, forming a basic slag that could later be detached from the metal.

This innovation, which was an immense success and had considerable economic implications,* had features both of the first industrial revolution (it was the indirect result of progress in other industries and the original idea came from a non-scientist) and of the second industrial revolution in which technology joins up with science. Without the basic scientific knowledge of Gilchrist (the chemist) regarding acids and bases, Thomas's idea could not have become an innovation. What might be regarded as a minor incident represented in reality a major change in the mechanism of the industrial system.

In the second half of the nineteenth century, technical progress became inseparable from scientific knowledge, as Landes indicates in the following passage:

> "Behind this kaleidoscope of change — sometimes marked by brilliant bursts, sometimes tedious in its complex fragmentation, always bewildering in its variety — one general trend is manifest: the ever-closer marriage of science and technology. We have already had occasion to observe the essential independence of these two activities during the Industrial Revolution; and to note that such stimulus and inspiration as did cross the gap went from technology toward science rather than the other way. Beginning in the middle of the nineteenth century, however, a close alliance develops; and if technology continued to pose fruitful problems for scientific, research, the autonomous flow of scientific discovery fed a widening stream of new techniques" (p. 323).

While the achievement of Thomas and Gilchrist in the history of steel provides an ideal example of the transition to the new technology, by 1878 the second industrial revolution had already been in progress for many years. The first revolution had hardly begun to reveal the full extent of its possibilities on the Continent before the first discoveries of the next wave started to flow from the laboratories.

These innovations enabled a new cluster of industries to enter the market. The start of the modern industry providing wood pulp for the manufacture of paper, for example, goes back to 1855. The discoveries of Volta, Ohm, Faraday, Siemens and others in the course of the nineteenth

* Between 1860 and 1890 the actual cost of producing steel fell by 80–90 per cent.

century paved the way for the use of electricity in manufacturing and in daily life after the building of the first power station in Britain in 1881. However, the biggest strides were made in the chemical industry with the introduction of the electrolytic process for extracting aluminium from bauxite (1866) and the Solvay soda process that replaced the Leblanc process by making use of ammonia from coal distillation at gas works (1863). This was also the period when organic products, such as benzene and dyes, were developed.

In the power sources sector, E. Lenoir developed the first practical version of an internal combustion engine, so opening the way for a major advance in public mobility.

Mechanization spread far and wide, as can be seen from the marketing of sewing machines and of a vast assortment of industrial machinery. The latter benefited from a whole series of improvements made at the different stages of steel production to eliminate bottlenecks when using the new, faster operating furnaces. Continuous rolling, for instance, made it possible to step up the supply of sheet steel, and quality improvements allowed precision engineering to develop.

All the innovations mentioned were the result of co-operation between laboratory researchers and engineers. In contrast to the preceding century, a "knack" for finding a way to increase productivity was no longer enough; it was a question of finding the best technical means for manufacturing products of the highest quality as cheaply as possible and in the largest possible quantities. The marriage between science and technology is evident, for example, in the production of new measuring instruments during the period, and in the advent of laboratories with consultants specializing in scientific and technical research for industrial purposes.

This, of course, raises the question why such a change occurred in this particular period. Two factors seem to have been of particular importance. One was the increasing difficulty of making further progress in improving industrial techniques. The other was the advance in education. By the end of the nineteenth century the simpler kinds of inventions had already been made. There was, of course, still scope for practical ingenuity, but only for marginal improvements and not for innovations of historic significance. For these, complex problems had to be solved, and the solution involved both knowledge of physical laws and the structure of matter and knowledge of production engineering.*

Success in this new field naturally depended on the availability of research workers and engineers with adequate theoretical knowledge. Hence, an advanced educational system was a major advantage in the race for industrial power that had begun. Being less well equipped in this field than the new Germany, Britain began at the end of the century to feel the first attacks on the economic supremacy that it had won through being the pioneer of industrial development.

Not all the innovations of the second wave of technical progress had immediate repercussions. In the first place, the progress of the second industrial revolution was more scattered over time and space than that of the eighteenth century innovations. Second, most of the countries of Europe were still in process of industrializing; Britain and Germany, and possibly France and the United States, were the only countries with the level of incomes and education needed for the marriage between science and technology. Third, the innovations in question were of a kind that inevitably involved a lag of several decades between the time when they were ready for use and the time when the necessary adjustments had been made in the areas of production, trade and attitudes. Certainly, the process began immediately, but it took a considerable time.

By the end of the century, however, the first indications of the profound changes brought about by the second industrial revolution had appeared. Again, these were visible in terms of economic growth, population growth, social organization and the general level of prices.

From 1870 to 1910 the population of Europe rose from 290 million to 435 million, or nearly 50 per cent. During the same period, national incomes doubled or tripled. The rise in *per capita* incomes produced the first hints of the mass consumption society. The fall in the relative prices of manufactured goods made incomes go further, so increasing the tendency for more of such goods to be included in normal consumption patterns. In addition, technical progress also began to affect the tertiary sector, i.e. transport, banking, commerce, etc. The distribution sector

---

*   It is worth noting also that technical progress in the nineteenth century had facilitated a marriage between science and technology. By providing scientists with suitable instruments for practical testing of hypotheses, technology had greatly helped physics and chemistry to complete the transition from a speculative science linked with philosophy to an experimental one.

also underwent a revolution at this period, with the development of department stores, chain store organizations and advertising.

The effect of all this was a change in attitudes towards consumption. In contrast to experience hitherto, an increasingly large part of the population was able to buy things that were not absolutely necessary, and the rural exodus and concentration of population in large urban centres promoted this attitude and made it a necessary one.

In spite of these developments — which would today be regarded as inflationary — the impact of the innovations of the first and second industrial revolutions on costs of production were such that prices collapsed in the period 1873—96. This major deflation merely reflected on a larger scale the general trend that had prevailed throughout the nineteenth century. It serves today as a measure of the extraordinary increase in productivity at the time. It has consequently been regarded as a barometer of industrialization in Europe. It was only in the very last days of the century that prices began to rise again, with beneficial effects on profits. Once again, the discovery of gold deposits (in South Africa and Australia) played a part in reversing the trend by increasing the amount of money in circulation.

The years between 1896 and 1914 are still remembered as the "Belle Epoque" in Parisian life, as the post-Victorian period and as the era of the Exposition Universelle in 1900, but to historians they are the period of the colonial and frontier skirmishes that foreshadowed the preparations for the Great War. In spite of the promises for human liberation implicit in the current changes in industrial technique,* energies were primarily directed to getting ready for the holocaust — whether one views this as a clash between sovereign nation-states or between opposed social classes.

It is true that the early stage of the second industrial revolution had not been accomplished without sacrifices. The period 1873—96 had left its mark on mental attitudes, not only because the deflation and decline in profits had undermined the belief in progress, but also because the doctrine of free trade was seriously shaken by the resurgence of protectionism and the formation of cartels to take advantage of the conditions created by the new technology.

---

* While the social progress made possible by the growth of productivity must be given its due, the second industrial revolution also opened the way to the "scientific management" movement. Where this forced the worker to keep up with the increased speed of the new machines and deprived him of all scope for judgment and inventiveness, it cannot be regarded as conducive to human liberation.

The fact that governments abandoned low duty policies or reverted to high tariffs in Austria (1874), Russia (1877), Germany (1879), Italy (1887), France (1892), etc., indicated Britain's loss of influence. The open defiance of the nation-states in the economic field was due to the success of the Industrial Revolution on the Continent, and the ending of British hegemony was one of the first consequences of that success. The parallel formation of cartels was not unconnected with the advent of the new science-linked technology. We have already noted that the technology was partly a response to a need to rationalize the production of innovations owing to their increasing complexity. But rationalization was not limited to research and development of new techniques. It was also applied to the organization of markets where the need to avoid "destructive" competition became urgent as investment costs grew with the greater complexity of techniques and scale of operation. It is particularly significant that the cartels formed at the end of the century were in growth-inducing industries: steel and chemicals. And they were most prevalent in Germany and German-speaking Switzerland where the second industrial revolution was most successful, and where firms arose that would still be among the giants of world industry a century later (BASF, Höchst, Agfa, CIBA, Geigy, etc.).

It is not surprising, therefore, that the end of the century coincided with growing pessimism. This was especially noticeable in literature and philosophy, and reached its fullest embodiment in the works of Oswald Spengler at the beginning of the twentieth century. The new mood reflected the anxieties created by structural change in a period of transition. Despite the maturation of the second industrial revolution in the second third of this century, the feeling of anxiety has never yet totally disappeared owing to the economic and social transformation brought about by science-based technology. The feeling has indeed been made more acute by the experience, in two World Wars, of the power of destruction that such technology can create.

In spite of two murderous wars and the Great Depression of the 1930s, there was no setback to economic growth over the period 1913–45 taken as a whole. The table below, taken – like the other figures quoted in the text – from a United Nations study [16], show that the rate of increase in production was lower in this period than in the preceding period, but that it continued to be on average above zero.*

Period:	Average annual growth rates of total production 1870–1913	1913–50
Country		
France	1.6	0.7
Germany (F.R.)	2.9	1.2
Italy	1.4	1.3
Netherlands	2.2	2.1
Sweden	3.0	2.2
United Kingdom	2.2	1.7
United States	4.3	2.9

This continuance of economic development must be attributed to the progress of the second industrial revolution. The process of technological change already begun was now entrenched in the system and altering it profoundly, and the political and financial disturbances were not enough to counteract the effect of this process altogether. Hence, productivity still rose between 1913 and 1938 by 1.3 to 3.0 per cent per year (according to the country considered), as compared with 1.5 to 2.4 per cent annually from 1870 to 1913. While some industries, such as textiles, metals and shipbuilding, were already facing a decline, others were starting the steepest part of their logistic curve of development and were inducing expansion in all areas of the economy. The inter-war period saw the development of the motor car industry, and the birth of electronics with a revolutionary new product, i.e. radio, leading to regular broadcasting in the United States and the Netherlands from 1920 and in Britain from 1922. The chemical industry expanded, especially in the area of artificial fibres where the pilot plant for nylon started by Dupont in 1938 was the fore-runner of the vast development of synthetic fibres after the Second World War.

It is worth noting that the inter-war period was also the time when macro-economic policies to sustain demand were first introduced. These were initially based on practical considerations: first, the high wage policy advocated by Henry Ford so that workers at his plants could afford to buy his cars; and later on, the programmes of major public works started in various countries (the New Deal in the United States and the motorways

* The averages disguise the wide variations in the chronological series, but underline the fact that the basic trend was nevertheless towards growth.

and rearmament programmes in Germany) to offset the effects of the depression. A theoretical basis for such policies was provided in 1936 with the publication of Keynes' General Theory [10] in which effective demand was the driving force for the whole economy. Keynes showed that, if a full employment equilibrium is to be reached, it is sometimes necessary for the State to influence aggregate demand — for example, by government spending — so as to change the pessimistic expectations of economic decision-makers and encourage them to invest. From now on, no reliance was placed on the theory by which the nineteenth century had been guided, namely, that the total value of the commodities produced always equals the total value of the commodities sold, so that there cannot be over-production leading to unemployment.

The change of attitude towards demand and government intervention in the economy no doubt had a good effect on growth during the inter-war period. Nor is there any doubt that it made a by no means insignificant contribution to the reconstruction of Europe and the economic success of the Western countries immediately after the Second World War.

All the same, the prospects seemed far from promising to economists at the end of the 1930s. We have already seen that the "stagnationists" were one of the main schools of thought at that time. And the first genuine model of economic growth, worked out at the same time by Harrod and Domar, was based on the hypothesis of a production function with set coefficients of labour and capital, i.e. further technical progress was not allowed for, and growth could well occur with under-employment of one of the factors of production.

Such pessimism is easy to understand if one considers the circumstances of the time. Europe had just passed through a succession of extremely serious monetary crises. The inflation that followed the 1914—18 war had shaken confidence in several countries, reaching catastrophic dimensions in the postwar German Republic; moreover, speculation on foreign exchange had precipitated the fall of the gold standard system, which was replaced in the 1930s by a system of flexible exchange rates. In all the countries persistent unemployment had set in well before the 1929 slump, reflecting the difficulties of certain declining industries in adapting to changed circumstances.* In addition, the return to protection at the end of the nineteenth century was not replaced by more liberal policies after

* See Landes [12], p. 369.

the war, as might have been expected. In fact, protectionism expanded to include Britain in 1921, and by 1927 the duties imposed by the European countries on consumption goods were on average 50 per cent higher than in 1913.

But the biggest cause for pessimism was the great slump in 1929, with the collapse of the banking system which brought down the whole production structure with it. The Great Depression of the 1930s was not, in fact, a basic change in the long-term growth trend, but an accident comparable with the overproduction crises in the nineteenth century: it differed only by being on a quite exceptional scale.

The crisis can be traced back to monetary causes – a reversal of the direction of the capital flows between the United States and Europe – followed by a chain of bank failures due to the collapse of the Kreditanstalt in Vienna. Unfortunately, the deflationary effect of the bankruptcies was reinforced by the wage reduction policy initially adopted by most of the governments. Far from reducing unemployment, the policy pushed it to even higher levels.

With such a catastrophe, underlining the interdependence of the nations that had been swept along by the industrial revolution, there was obviously little to encourage optimism. Yet actually the world was on the verge of a period of unprecedented economic growth.

A series of articles ("The U.S. Frontier") published in *Fortune* in 1939 shows that opinions differed among economists on the subject of growth. The business world was more aware of the changes that had taken place in the previous decades in the relationship between science and technology. It had a feeling that the phenomenal development of industrial research (the number of persons employed in this field in the United States rose from 8000 in 1920 to 17,000 in 1927, and to 42,000 in 1938) could not leave the economy unaffected. And indeed the results in terms of economic growth, as shown in the table below, fully justified their optimism.

It is true that, this time, the situation was not one that could hamper growth: the Second World War, unlike the First, had eliminated a number of barriers. In the first place, maximum military use had been made of all that science and technology could offer – radar, atomic energy, jet propulsion, operations research and system analysis – thus providing a set of techniques which the industrial system could benefit from, and which led among other things to the emergence of new industries, such as data processing and aerospace.

	Average annual rates of growth of Gross National Product (percentages)	
Country	1950–1960	1960–1970
France	4.4	5.5
Germany (F.R.)	7.6	5.1
Italy	5.9	5.5
Netherlands	4.9	5.5
Sweden	3.3	4.5
United Kingdom	2.6	2.9
United States	3.2	4.6
Switzerland	4.0	4.4

Secondly, protection was under attack at various political levels. There were the Bretton Woods agreements on international payments, and the General Agreement on Tariffs and Trade designed to eliminate trade barriers. At the European level, the European Payments Union and the Marshall Aid programme helped to reorganize and relaunch the economy, while avoiding the disastrous reparations policy that followed the First World War.

The effects of this rehabilitation of the Western economy are well known: a rapid rise in the standard of living, further industrialization and urbanization, and internationalization of the economies. The two tables below bring out two things, i.e. the increasing share of foreign trade in the overall economic activity of Western Europe, and the increasing importance of industry at the expense of agriculture. It should be noted on this last point, however, that more recently the proportion of persons working in manufacturing has fallen, with a corresponding increase in the services sector.

Average annual growth rates of major aggregates in the national accounts of Western Europe, 1950–1968 (per cent)	
National income	4.6
Consumption	4.5
Public expenditure	3.4
Gross fixed capital formation	6.7
Exports	7.6
Imports	7.8

Sector	Share of major sectors in the GNP of Western Europe 1950	1968
Agriculture	10.5	7.3
Manufacturing	34.0	39.1
Others	55.5	53.6

What is remarkable from the point of view of economic history is that at no time during this period of almost three decades was there a drop in the chronological series of annual national income statistics. The rate of growth in GNP fluctuated from one year to another but until the depression in 1974 had never been negative.

Naturally, such a long period of rapid growth has inevitably affected ways of living, and also ways of looking at economic problems. Reliance on growth to solve all resource allocation difficulties and all social tensions has become a mental reflex. At the end of the 1960s and beginning of the 1970s, two-figure growth rates held the same attraction for politicians as Eldorado did for their ancestors.

Obviously, a short-term decline in business activity involving a drop from 6 per cent to 2 per cent in the rate of annual growth was felt as a serious economic setback, threatening political and social upheavals.

The constant watch kept on everything related to growth encouraged economists to investigate the causes of growth. The studies quickly showed that growth was not primarily due to an increase in capital and/or labour but to a "residual" factor. In ordinary language, this meant an increase in the productivity of capital and labour, i.e. it measures the impact of technological progress.

Share of the "residual" factor in the growth of certain countries, 1950—62 (per cent)			
United Kingdom	52	Denmark	56
Italy	70	France	75
Japan	60	Germany (F.R.)	62
United States	41	Netherlands	60
Belgium	63	Norway	70

Realization of the importance of technological progress and of the link between this and scientific research gave rise to two firmly held beliefs. First, a very sweeping belief that, thanks to the new technology, humanity was at last on the way to the era of plenty and leisure for all. Second, a conviction that this could be reached by "simply" increasing the proportion of expenditure on research and development. This vision of the world — exemplified in the mid-1960s by Jean-Jacques Servan-Schreiber's book, *The American Challenge* — gave further impetus to scientific and technical research. In 1970, there were two million people employed in research in Western Europe, or four times more than in 1950. And the fact that the leading industries from the technological point of view were those with the highest rates of business growth encouraged industrialists to follow suit. The table below shows that governments did not hold themselves aloof in this area, and that the share of R & D in the rising national budgets was on the increase.

Country	Average annual growth rates of national budgets and government expenditures on R & D, 1963–1967 (per cent)		
	Budget	R & D expenditure	Share of budget
France	9.5	14.0	2.7
Germany (F.R.)	8.5	18.0	1.5
Italy	11.0	13.1	0.6
Japan	15.0	19.7	1.6
Netherlands	14.0	14.9	2.0
United Kingdom	10.4	5.8	3.5
United States	8.9	1.2	7.8

The effect of the down-turn in 1975 was all the harsher because of this. Yet for ten years or more, growing awareness of the adverse effects of growth had been preparing public opinion for a cessation of growth. It was being pointed out at the international level that the rich were getting richer, while the Third World countries were dropping behind. It looked as if the industrialized countries were taking advantage of the economic weakness and political dependence of the developing countries to appropriate their raw materials, sources of energy and labour power. At the national level, attention was being drawn to the damage to the environment, the disadvantages of the latest technology, and the costs of growth in general.* The Club of Rome's report on limits to growth in 1972 came only two years before growth actually stopped. There had been a number of warnings of the break. Apart from the crisis of confidence in the West, as shown by the 1968 students' revolt, persistent inflation, the

* Cf. in particular Rachel Carson [2], E. J. Mishan [15], G. R. Taylor [19] and P. and A. Ehrlich [4].

malfunctioning of the international monetary system, a revival of protectionism, and the weakening of United States economic and political influence, were all premonitory symptoms or direct causes of the breakdown.

2.3. A third industrial revolution or the limits to technology?

The historical developments recalled in the foregoing pages do not represent in any way an original view of economic development over the past 200 years. It is very widely accepted today that economic growth in the modern era was made possible by the progress in technology. It is worth noting, however, that progress was not constant during the period considered. It could only be maintained and broadened because of two waves of technological discoveries. Each wave was in reality a family of discoveries that became technical innovations when taken over by new industries. The discoveries spread gradually through the whole economy, but caused profound structural changes while doing so.

The first wave, starting in Britain in the mid-eighteenth century, had its initial impact on the textile industry, before affecting metallurgy and transport and then dying out at the end of the nineteenth century — after having spread to Continental Europe and North America.

The second wave, characterized by the union of technology and science, took shape from the end of the nineteenth century and has continued to spread up to the present, affecting the entire world.

In many ways, the period through which Western society is now passing might be compared with the end of the nineteenth century. At that time, the economic future looked dark, and this was reflected in the return to protection and in persistent deflation. The antagonisms that sprang up with the relative decline of Britain prefigured those now emerging with the weakening of American leadership. Yet it was, in fact, only a relatively brief period of transition between the era of the first industrial revolution and that of the second.

Should the present period also be regarded as a transitional stage on the way to a fresh era of economic development? Are we on the verge of a third industrial revolution, equally radical in its effects and directed in the same degree towards maximizing the economic power and wealth of the nations?

Few would be bold enough to rule out such an outcome. It is quite conceivable that, if all the potentialities of informatics and nuclear energy are exploited and there are breakthroughs in the development of less accessible resources (e.g. in the oceans), this will lead to new growth, to a new stage on the way to a world of plenty, and to new changes in society. The present tensions would then be merely a symptom of the difficulties that occur when any system is changing.

One must point out, however, that this would not strictly speaking be a third industrial revolution but rather a "second wind" of the preceding one. The linkage between science and technology would simply be reinforced, and this would — in the minds of the upholders of this view of the future* — be enough to prolong past economic growth.

Historical comparisons may be made to yield a different interpretation of the present depression. Rome in its decline, the empire of the Incas and even medieval France** are often quoted as a basis for predictions of a durable decline of the West and of rationalist thought. Human history and the plethora of information about the world in which we are living provide an inexhaustible mine for anyone seeking to find parallels for selected aspects of past and present reality.

At the opposite extreme from such an all-embracing viewpoint are those who will not admit that the depression in the Western countries has underlying causes and is linked up with the evolution of civilization as a whole. In their view, the present economic troubles are due to errors in the short-term conduct of public and private business. This is no doubt true if one is thinking of the *immediate* cause, and it would not occur to us to deny that there is a relation between the lack of policy in monetary questions and the "stagflation" that has overtaken most of the world's countries in the mid-1970s. But from the wider angle of the evolution of the whole system, it tells us nothing about the underlying reasons for the mistakes: Why were the world growth of international payments facilities and the dismantling of the international monetary system allowed to happen, when it could already be foreseen (and Cassandras were not lacking) that the policy would have disastrous results?

The possibility that present economic difficulties are transitory and only an accident in a general line of growth cannot of course be entirely

* On this point, see the books by Hermann Kahn.
** See the book by J. Gimpel [7].

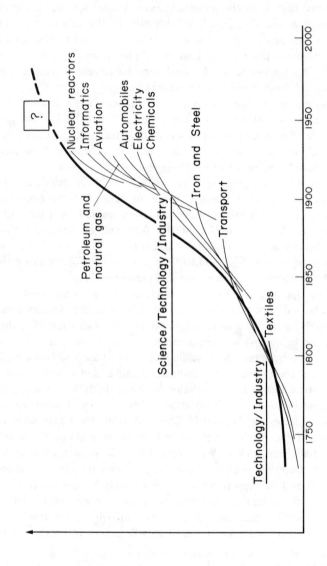

Fig. 1. A schematic view of the effect of the technological revolution on global economic growth through successive sectoral advances.

ruled out. But there are many indications that weaken this argument.

The following chapters are an attempt to set out in an orderly manner certain scattered elements of the factual situation, which — when taken together — make it clear that the historical conditions explaining growth in recent decades no longer exist. If there *is* a third industrial revolution, it will not be mainly due to a reinforcement of the linkage between science and technology. The technology of the first industrial revolution reached its limits at the end of the nineteenth century. It looks as if the alliance between science and technology that was characteristic of the second revolution is reaching its limits today.

Our argument is that, in a given historical context, the embodiment of technical progress in the production sphere made a period of high growth possible. But the impact of this new factor has gradually declined, until today it is negligible. In other words, technology has operated like any other factor of production employed in increasing quantities: after an extraordinarily successful period, later extended by the injection of science, it has recently reached a phase of declining returns (see Fig. 1). The *global* marginal productivity of any investment in R & D is today tending towards zero, which explains the slowdown and stoppage of growth.

This general outline fits into a structural view of the present crisis. It provides certain points for reflection on the reality of what is happening today. It does not exclude the possibility of a further period of prosperity for Western society — perhaps as a result of innovations that provide the foundations for an economic and social equilibrium different from the one we have been striving to attain up to now.

References

1. Paul Bairoch. *Révolution Industrielle et Sous-Développement.* Soc. d'Ed. d'Enst. Sup., Paris, 1963.
2. Rachel Carson. *The Silent Spring.* Houghton Mifflin, Boston, 1962.
3. James Duesenberry. *Income, Saving and the Theory of Consumer Behaviour.* Harvard University Press, 1949.
4. Paul and Anne Ehrlich. *Population, Resources, Environment.* Freeman & Co., San Francisco, 1970.

5. Claude Fohlen. *Qu'est-ce que la Révolution Industrielle?* Robert Laffont, Paris, 1971.
6. Jay Forrester. *World Dynamics.* Wright-Allen-Press, Cambridge, Mass., 1971.
7. Jean Gimpel. *La Révolution Industrielle au Moyen-Age.* Seuil, Paris, 1977.
8. Olivier de La Grandville. *Théorie de la Croissance Economique.* Masson, Paris, 1977.
9. Gottfried Haberler. *Prosperity and Depression.* Atheneum, New York, 1963 (first printed Geneva, 1943).
10. John Maynard Keynes. *The General Theory of Employment, Interest and Money.* Harcourt Brace, New York, 1936.
11. Simon Kuznets. *Economic Growth and Structure.* Heinemann London, 1966.
12. David Landes. *The Unbound Prometheus.* Cambridge University Press, 1972.
13. Dennis Meadows *et al. The Limits to Growth.* Universe Books, New York, 1972.
14. John Stuart Mill. *Principles of Political Economy.* University of Toronto Press, 1965 (1848).
15. Ezra Mishan. *The Costs of Economic Growth.* Pelican Books, London, 1967.
16. United Nations. *The European Economy from the 1950s to the 1970s.* United Nations, New York, E. 72. II. E.1., 1972.
17. Walter Rostow. *The Stages of Economic Growth.* Cambridge University Press, 1967.
18. Bertrand Russel. *The Scientific Outlook.* London, 1931.
19. Gordon Rattray Taylor. *The Doomsday Book.* Thames & Hudson, London, 1970.

CHAPTER 3

Technology and the Production Function

The production function is one of the fundamental relationships in economic theory. It indicates how a number of variables representing the "factors of production", i.e. goods and services used in production, are related to the quantity of goods or services produced.

Mathematically, this is a function with more than one independent variable:

$$q = f(x_1, x_2, x_3, \ldots) \tag{1}$$

where x_1, x_2, x_3, ... represent the factors of production, or inputs, and q represents the quantity, or output, produced per unit of time.

Equation (1) may refer to a firm (micro-economic) or to an economy as a whole (macro-economic).

In the first case, it indicates the quantity of a homogenous good produced by a particular firm, e.g. the number of pins manufactured in a day. The number and nature of inputs will accordingly vary with the type of production. The harvesting of walnuts in the Perigord district of Southern France, for example, requires labour but comparatively little capital. If 100 pounds of nuts are gathered in an afternoon by three men beating the trees for 4 hours, the inputs involved are an orchard of walnut trees, 12 hours of labour and three stout poles. In contrast the labour input in some of today's automated plants is very small.

In the second case, equation (1) would indicate the total amount of goods and services of different kinds produced during a given period in the national economy as a whole. Here, all the available factors of production are represented according to their contributions to the global output. It should be pointed out, however, that economists, when constructing their production functions, have always been strongly influenced by actual conditions at the time when they were living. In Ricardo's time, agriculture was still the predominant form of economic activity, though there was also a promising manufacturing sector. Hence, land ranked first among the factors of production, followed by capital and labour. The Classical production function is generally of the following pattern:

$$q = f(T, K, L) \qquad (2)$$

where T represents land, K capital and L labour.

When there was a major deflation at the end of the nineteenth century, interest rates fell so low that some economists proposed a production function with only land and labour as factors, treating capital as a more or less "free" good available in unlimited quantities. Since that time, capital has become so much more significant than land, not only in industry but also in agriculture, that land has disappeared from the Neo-Classical production functions, e.g. the Cobb-Douglas and C.E.S. functions, being subsumed under "capital". The usual production function thus becomes:

$$q = f(K, L). \qquad (3)$$

However, some writers have tried to set up more detailed relationships. The production function suggested by Georgescu-Roegen [4] comes into this category. It makes a distinction between the usual production factors (labour, land and capital) and other factors normally associated with capital such as natural resources (including "free" goods such as air and water), intermediate goods (materials) and products used for maintenance. The first group represents utilization of a *"fund* of services" and the second represents consumption of a *flow* of products.* The output of the system comprises the flow of final products and the pollutants discharged by the production process.

If the "dynamic" aspect of the system is also covered by making each independent variable a function of time and no longer an ordered sequence of real numbers, the following equation is obtained:

$$Q(t) + W(t) = \mathscr{F}\left\{H(t), L(t), K(t); R(t), I(t), M(t)\right\} \qquad (4)$$

where $0 \leqslant t \leqslant T$ represents time, and the symbols have the following meaning:

---

* Georgescu-Roegen's analysis could be further refined by classifying renewable natural resources (air, water, materials derived from plants or animals) under the fund of "services" rather than the flow of products.

Q: output   H: labour
W: pollution   L: land
                 K: capital
                 R: natural resources
                 I: intermediate products
                 M: maintenance products;

Equation (4) reveals the entropic nature of the production process. On the left-hand side, $Q(t)$ represents a product flow to be consumed and transformed finally into waste. The ultimate effect of the production system is therefore to transform into waste a flow from the planet's stock of raw materials and energy, using up a fund of service-generating factors. Georgescu-Roegen sees a close analogy between the economic process and the Second Law of Thermodynamics.* The difference lies in the fact that entropy in the universe is "blind" and without purpose, whereas the entropy of the economic process — which represents both a part of and an acceleration of the entropy of the universe — has a *raison d'être*, i.e. human enjoyment of life.

We shall have occasion to return later to the entropy of the economic system that is evident from equation (4). However, while the apparent comprehensiveness of the equation makes it attractive, we feel that there is one serious omission. It includes no indication of the influence of technology on the production system. This gap is surprising, since the recent history of economic theory records many attempts to introduce this factor into the production function, all of them in response to the evidence of a "residual factor" in economic growth.

The role of this factor was confirmed by R. M. Solow [7], who found that one-third of the growth in the American economy in the recent period was attributable to capital, and two-thirds to technical progress. Similarly, at the micro-economic level, Freeman's study [3] for the period 1935—58 in the United States and Britain showed a positive correlation between growth in the manufacturing sector and the investment in R & D.

---

* The first two laws of thermodynamics are: (1) The energy of the universe remains constant (Law of the Conservation of Matter and Energy); (2) The entropy of the universe moves at all times towards a maximum (Law of Entropy). The Law of Entropy means that the energy in the universe is continually undergoing a qualitative change from a state of "availability" (concentrated, usable) to a state in which it is dissipated and non-recoverable (entropy).

Nevertheless, apart from a few isolated attempts (e.g. by Bensoussan [1]), technology has not been included in the production function as a genuine factor but as a coefficient of the existing factors. The most controversial point has been whether technical progress makes it possible to economize capital or labour, or both. Accordingly, the coefficient was applied to capital, to labour or to the whole function.

There are two reasons why this approach was chosen. One is that technology was seen by economists as a phenomenon lying outside the field of economics. It could only appear in the models, therefore, as a datum and not as a separate variable. The second is that Classical economic analysis had shown that the law of diminishing returns to factors of production has a profound effect on the global equilibrium, and it did not seem possible on *a priori* grounds to apply the law to technology. Indeed, it looked as if technology could suspend the law of diminishing returns, or at least push back the limits that it imposed on the growth of the production system.

The law of diminishing returns indicates that, where there are successive increases of one factor of production, with the other factors remaining at the same level, the corresponding increases in output become smaller. For example, as the number of workers in a workshop is increased without any change in the number of machines, production increases less and less* since the shortage of machines forces the additional workers to be idle part of the time. In the medium term, of course, the quantities of the different inputs can be increased simultaneously, but this is not possible indefinitely as there are limits to the expansion of a factory.

Since Ricardo systematized the law and made it the basis of his theory of rent, it has been an essential element in economic theory. Its best known application is perhaps Malthus's law of population, which indicates that agricultural production cannot keep up with population growth since land is a limited factor. But it is worth noting, without going into the details, that the law is used everywhere – in partial equilibrium analysis (product markets, factor markets) and in Neo-Classical or Keynesian general equilibrium theory – and its influence often makes the attainment of stable equilibria easier.

Nevertheless, it is not a specifically economic law. It applies in all areas where effort is demanded, including the intellectual field and the field of sport. A tennis player's output is lower in the fifth set than it was in the

---

* Or more precisely, marginal production falls.

first. And the law applies, though expressed differently, in other sciences.

In system dynamics, it takes the form of negative feedback loops that prevent the system from pursuing a runaway course of exponential growth (see Forrester [2]).

In physics, it is implicit in thermodynamics. As is well known, the origin of this branch of physics was Carnot's book in the early nineteenth century on the efficiency of steam engines. Every heat engine operates through a transfer of heat from a hot to a cold body, and only part of the quantity of heat energy employed is used in the work performed by the engine. The efficiency of the engine is given by the following formula:

$$E_t = \frac{Q_1 - Q_2}{Q_1} = 1 - \frac{Q_2}{Q_1}$$

where Q_1 represents the quantity of heat energy derived from the hot source (i.e. an increase in entropy) and Q_2 the heat energy conveyed to the cold body during time t. The efficiency is less than unity and can be raised only by a more than proportional increase in the energy consumed – hence the affinity with the law of diminishing returns.*

At the macro-economic level, the law of diminishing returns is an obstacle to continuous growth in the national product. This was recognized by the Classical Economists, and it was the reason for their belief that a stationary state was inevitable in the longer term. Some of them (including Ricardo) indicated the possibility that technical progress could delay the arrival of the stationary state, but none could have foreseen the fantastic development of science and technology during the last century.

Impressed by the way in which technology has rendered a good deal of the earlier theories obsolete, later economists have not unnaturally looked upon it mainly as an antidote to the law of diminishing returns. Most have treated it as a multiplier of the effects of the other factors of production. Others defined it in such a way that it could be a source of exponential growth in models, i.e. they made it an exponential.

* From:
$$E = \frac{Q_1 - Q_2}{Q_1}$$

the following can be derived:
$$\frac{\delta E}{\delta Q_1} = \frac{Q_2}{Q_1^2} > 0 \text{ and } \frac{\delta^2 E}{\delta Q_1^2} = -\frac{2Q_2}{Q_1^3} < 0.$$

In other words, the acceleration of entropy does not prevent the increase in output from declining.

function of the national product so that, in the last analysis, technology induces technology, at an accelerating pace.

It may be wondered if recent writers have not, like their predecessors, suffered from a certain one-sidedness. The latter put forward the law of diminishing returns because the society they knew was pre-industrial or just beginning to industrialize. The former have neglected diminishing returns because they are surrounded by a spate of technological achievements. But have they analyzed the technological phenomenon as a whole and have they studied the causes of the scientific and technological revolution and the direction in which it is moving?

Let us say straight away that it no longer seems reasonable to consider any production function in which technology is not one of the factors of production. In the real world of contemporary industry, there are many branches in which expenditure on R & D represents a sizeable percentage of total production expenditure: for example, aviation, electronics, chemicals (especially pharmaceuticals), petroleum and natural gas, etc. To treat technology as one of the factors of production would simply be a recognition of the facts.

Some economists object that technology is merely the output of a combination of capital and labour, and that these two factors alone should appear in the production factor.

This objection does not stand up to close examination. Technology is not a mere function of capital and labour. It is the result of the complex interaction of different elements, in which time and uncertainty play a considerable part.

Firstly, while it is true that capital and labour are needed for producing technology, the labour and capital in any given period are not being used to generate the technological innovations actually employed in that period. They are helping to produce innovations that will be used to obtain the output of future periods.

The technology employed today is the fund of knowledge accumulated in past periods.* In other words, it is a function of the capital and labour

* To define technology as a fund of knowledge may seem an over-simplification, but this definition is sufficient for our present purpose. The few economists who have attempted to analyze technology make a distinction between technological knowledge in the strict sense — i.e. technical and scientific knowledge relating to production and management — and human know-how (see Matthews [6]). In any case, it is technology and not simply technical progress that must be included in the production function: technology is a production factor, technical progress a factor in economic growth.

employed in past periods.

For this reason, technology could not be reduced to a function of capital and labour simply by proceeding from the production function:

$$Q = f(K, L, Z) \qquad (5)$$

where K, L and Z represent capital, labour and technology, to the function:

$$Q = F(K, L) \qquad (6)$$

on the basis of a function $Z = Z(K, L)$.

The time factor would have to be explicitly introduced into this technology function, giving:

$$Z_T = \sum_{t=-\infty}^{T-1} g_t \, (L^*_t, K^*_t) \qquad (7)$$

where Z_T, or the fund of technological knowledge used in the current period, consists of the sum of the technological innovations devised since the beginning of time, on the assumption that all of these have been handed down from generation to generation. L^*_t and K^*_t represent the fund of capital and of labour employed on the production of technological innovations during period t.

Consequently, it is no longer possible to reduce equation (5) to equation (6). Equation (5) must be rewritten as:

$$Q_T = f \left\{ \sum_{t=-\infty}^{T-1} g_t(L^*_t, K^*_t), K_T, L_T \right\}. \qquad (8)$$

Thus, a dynamic type of production function is required.

Secondly, technological innovation is not only a function of capital and labour. Equation (7) gives only an incomplete account of the way in which technology is generated.

To analyze this process, it is necessary to make a distinction between technological innovation, or the industrial application of a scientific and technical discovery through a process of R&D; and invention, i.e. an inspired and more or less accidental discovery of a new property, production or process.

An invention certainly involves the employment of capital and labour, but in modern conditions inventions are not often made independently of the stock of scientific and technical knowledge — and, more particularly, of changes in this stock and of the interaction of discoveries in different scientific fields. Nor can the element of chance in inventions be overlooked. Quite a short conversation with the head of R & D in a major company will show that inventions are for the most part a random variable. Out of a hundred research projects, perhaps ten, twenty or thirty will lead to usable results; the others will be abandoned or will produce results of no value. This is the reason why research managers are very much concerned with the variance of the success rate and any change in the mathematical expectation of this rate over time.

To take account of these factors, the flow of inventions at period t would have to be specified as a function:

$$iN_t = iN_t\,(\widetilde{K}_t, \widetilde{L}_t, S_t, \Delta S, u) \qquad (9)$$

where $\widetilde{K}_t$ and $\widetilde{L}_t$ represent capital and labour assigned to applied research during period t; S_t represents the stock of scientific and technical knowledge in t; ΔS the change in this stock, and u the random factor.

Starting with this flow of inventions, it is possible to make the transition to innovation* by employing a fraction of the available capital and labour for a certain time. This last point is extremely important. It would be unrealistic to picture the transition from invention to innovation as a momentary phenomenon within the confines of a given period. The process takes time and — as we shall see later — the trend in the time-lag between invention and innovation is a cause of anxiety in the technological field today, as is the trend in the research success rate.

To simplify the notation, we will assume that there have been no inventions prior to period 1, so the flow of innovations in any period T is the sum of the inventions in periods $T,\ T{-}1,\ T{-}2,\ \dots,\ t,\ \dots,\ 2,\ 1$, whose development ends at this date.

If $a_T^t{\cdot}iN_t$ is the number of inventions of the period t that are responsible for an innovation at period T, we have:

$$\Delta Z_T = a_T^T\,.iN_T + a_T^{T-1}{\cdot}iN_{T-1} + \dots + a_T^t{\cdot}iN_t + \dots + a_T^1{\cdot}iN_1$$

$$\Delta Z_T = \sum_{t=1}^{T} a_T^t \cdot iN_t. \qquad (10)$$

* It will be noted that we follow the familiar sequence of basic research, applied research and development (see [5]), with: change in scientific and technical knowledge (ΔS), invention (iN), and technological innovation (ΔZ).

The flow of innovations ΔZ is thus the endogenous variable of a distributed lag equation. The pattern of the lags is given by the distribution curve of a^t. The latter is not constant. Its shape depends on random factors (v) and on the capital and labour allocated to development during the periods 1 to T. Consequently, there is a different a^t distribution curve for each period T.

$$a_T^t = a^t \left[\overset{T}{\underset{1}{K^*}}(t), \overset{T}{\underset{1}{L^*}}(t), v \right]. \tag{11}$$

Combining (11) and (10), we obtain:

$$\Delta Z_T = \sum_{t=1}^{T} a^t \left[\overset{T}{\underset{1}{K^*}}(t), \overset{T}{\underset{1}{L^*}}(t), v \right] . iN_t \tag{10'}$$

and using (9):

$$\Delta Z_T = \Delta Z_T \left[\overset{T}{\underset{1}{K^*}}(t), \overset{T}{\underset{1}{L^*}}(t), \overset{T}{\underset{1}{K}}(t), \overset{T}{\underset{1}{L}}(t), \overset{T}{\underset{1}{S}}(t), \overset{T}{\underset{1}{\Delta S}}(t), w \right]. \tag{10''}$$

The flow of innovations at period T is a function of the capital and labour allocated to research and development in the earlier periods, of the stock of scientific and technical knowledge and its change during these periods, and of random factors covered by w.

As regards the fund of technology at period T, this is the sum of the flow of innovations in the periods preceding T:

$$Z_T = \sum_{t=-\infty}^{T} \Delta Z_t$$

It is now clear that — having regard to the peculiar nature of the technological factor — it would be totally inaccurate to treat it as a function of current capital and labour on the plea that technology represents accumulated capital and labour.

Moreover, capital itself is accumulated labour, but no one would seriously think of eliminating it from the production function. Otherwise, it might be argued by going back to Adam that the only two genuine factors of production are land and the divine breath of life.

Capital and labour are, of course, involved in the production of technology but in a time-based manner, and two other factors are involved as well: first, the change in the stock of scientific and technical knowledge – which means that technological progress is partly self-induced – and second, uncertainty. The latter covers both the element of pure chance that often accompanies discoveries and a host of cultural, psychological, sociological and political factors whose interaction helps to provide more or less favourable conditions for the production of technological innovations.

So, from equations (4) and (5) it is possible to set up a complete production function:

$$Q(t) + W(t) = \mathscr{F}\Big\{R(t), K(t), L(t), Z(t)\Big\} \qquad (12)$$

in which Q is output, W pollution, R natural resources and land, K capital, L labour and Z technology. In contrast to equation (4), intermediate goods and maintenance products have been omitted as factors of production since they appear both as outputs and inputs* in the macro-economic equation. But now technology has been included.

It should now be clear that technology is an autonomous factor of production. This leads to the question: is it reasonable to assume that the law of diminishing returns applies to technology in the same way as to labour and capital?

There appears to be no *a priori* reason for technology being immune. The reason why it was for long regarded as an obstacle to the operation of the law of diminishing returns is that it was in the process of emerging as a new factor in the production sphere. This new factor was pushing back the boundaries of the system comprising the older factors alone. This was quite natural: if capital (tools and animals) is introduced in a production system in which the only available factors have been land and human labour, it is to be expected that the productivity of these will rise steeply and that output will increase substantially. In the case of technology, the effect was all the greater because the emergence of the new factor brought with it a promise of further innovations: it freed capital and labour for developing innovations and created a favourable environment for further

---

* Nevertheless, imported intermediate goods and maintenance products must appear in the macro-economic production function.

scientific and technical discoveries as a result of improved living conditions and progress in science and education.

Paradoxical though it may seem, technological progress is in fact more vulnerable in the long run to diminishing returns than are labour and capital.

On the one hand, it may be affected by processes similar to those which lead eventually to a decline in the marginal productivity of capital and labour. In a world where usable land is scarce and where the fund of capital and the fund of labour each has its own tempo of growth, increased technological sophistication can in the long run only lead to marginal improvements. And the marginality of the improvements is likely to be all the greater because technical progress is also a source of distortions in the system.

On the other hand, we have seen that the passage of time, uncertainty and cultural factors, have an important bearing on the production of innovations. This means that the flow of innovations cannot be speeded up in the same way as an investment flow, by diverting production factors away from the consumer goods industries and towards the capital goods sector. In the case of technology, the allocation of a larger share of available resources to research and to development does not mean that there will be more innovations, if the inventions rate is simultaneously falling and the average time-lag between an invention and the corresponding innovation is increasing.

So the diminishing returns of technology are likely to be all the more noticeable where technological progress is slowing down and its productivity is declining.

In this case, the prospect is one of a slowdown or even a cessation of growth because the mainspring of growth is broken. Growth is not being held back by its external limits, such as the exhaustion of natural resources or pollution on a catastrophic scale. Long before such disasters, internal limits to growth become apparent and impose a more sensible socio-economic system, at the cost of tensions that will be all the more severe in that they were not foreseen.*

* Foreknowledge of the external limits would certainly have some influence on the emergence of the internal limits. Fear of an ecological catastrophe might well be a factor in a slowdown in technical progress leading to the abandonment of work on the development of new techniques, e.g. in the nuclear field.

In the following pages, we shall consider the factors which might suggest that such a process is occurring in the advanced industrial societies.

References

1. Claude Bensoussan. *Progrès Technique et Distorsions Economiques Internes?* Cujas, Paris, 1971.
2. Jay Forrester. *Principles of Systems.* Wright Allen Press, Cambridge (Mass.), 1968.
3. Freeman and Young. *The Research and Development Effort in Western Europe, North America and the Soviet Union,* 1965.
4. Nicholas Georgescu-Roegen. *The Entropy Law and the Economic Process.* Harvard University Press, 1971.
5. Orio Giarini. *L'Europe et l'Espace.* CRE, Lausanne, 1968.
6. Robert Matthews. "The contribution of science and technology to economic development", in *Science and Technology in Economic Growth* (B. R. Williams, Ed.). MacMillan, London, 1973, pp. 1–31.
7. Robert Solow. "Technical change and the aggregate production function", *Review of Economics and Statistics,* August 1957, pp. 312–320.

CHAPTER 4

Is Technological Innovation
Slowing Down?

Chapter 2 made the point that technology has not in the past progressed at a uniform pace, but in a succession of cycles associated with the rise and decline of particular industries. These have been long-term cycles, covering a succession of short-term cycles within each industry. The man-made fibres industry is a good illustration of this.

In the present century there have been two major cycles of technical progress in the textile fibres industry. The first goes back to the development of artificial fibres. The underlying idea here was that, since all natural fibres (e.g. cotton) consist to a considerable extent of cellulose, it might be possible to make use of cellulosic matter in other vegetable products, transforming it into fibres. Three or four viable solutions were found in succession, each leading to a development of the fibres industry up to the Second World War.

The second cycle was based on the idea that cellulose might not be necessary, and that some other chemical material with suitable properties might be substituted. The answer in this case depended on an advance in scientific knowledge concerning the structure of matter, since a more radical transformation was involved. It was no longer a question of modifying cellulose from plant life, but of taking a quite different raw material (petroleum) and changing its basic structure. In the era of synthetic fibres which now began, four families were developed in turn at varying intervals: the polyamides, then the polyesters and polyacrylics, and finally polypropylene. There were other families as well, but these were of lesser importance. On the other hand, the four major groups of synthetic fibres set off a boom in the chemical industry and enabled a small number of companies to become outstandingly successful.

The second cycle reached maturity in the 1960s. Since that time, the practical possibilities opened up by successive inventions have been increasingly marginal, and hopes of a "second wind" have been disappointed. In the end, research came to be mainly directed towards the improvement of existing methods, as there was no further scope for exploring the possibilities offered by science for modifying Matter. Research on the structure of the atom has still a long way to go before it can be of practical significance in most fields, and all the more so since the structure of the universe as a whole is not yet clear. True, an attempt has been made to move into a new cycle by finding a way of manufacturing textiles without going through the fibre stage, i.e. by finding a compromise between traditional textiles and paper. The idea was a good one and it produced non-woven materials, but unfortunately, the new revolution fell flat for two reasons. Firstly, it proved impossible to reproduce in non-woven materials the elasticity and porousness of woven and knitted materials. Secondly, the non-woven materials were planned in the anticipation that the mass consumption society would increasingly use disposable products. Now, feasibility studies showed that the advantage of eliminating laundering of textiles would be largely offset by problems in storing, marketing and disposing of products in such quantities. Hence, only one product of this type has had some success: a tufted carpeting in which some of the properties of felt are reproduced by incorporating 40–60 per cent of glue.

In the immediate future (up to 1985) it is highly improbable that there will be a new period of major innovations in this industry. The present struggle is for improvements and, while these are not without significance, they are not comparable with the advances made in the past. It can, of course, be argued that one cannot predict fundamental breakthroughs (e.g. a new fibre with all the characteristics of cotton). But even if this occurs, allowance has still to be made for the time-lag before it can be marketed. On a more pessimistic view, it may be that this sector has now exhausted the potentialities that it acquired at the end of the nineteenth century from the marriage between science and technology.

The question now is whether the dead-end in the man-made textiles industry is typical of what is happening in manufacturing as a whole, or whether it simply represents the ending of one line in a family of industries where each new generation, as it develops, will continue to sustain the momentum of economic progress in the West. There are still many sectors in which innovation is flourishing – e.g. electronics – and there are others where the scope and urgent need for further research may lead to great developments (biology, biochemistry, medical research, alternative sources of energy). On the other hand, there are also a great many items in the daily flood of information which suggest that a reversal of past trends is occurring, or at least that innovation is not progressing smoothly.

In the large computers sector, for example, IBM announced in March 1975 that it was ceasing work on the innovations planned under its Future Systems Program, and that this had been postponed for seven or eight years for various technical and business reasons. Rather than attempting a new technological breakthrough, the firm would devote itself to improving existing systems. A journalist commented in the *New Scientist* that this was almost as if the Vatican were to abandon the celibacy rule for priests!

A decline in the frequency of new products is expected in the chemical industry*, and even industries believed to be very strong on R & D are having difficulty in maintaining their rate of innovation (cf. Hurst's article [3] concerning electronics).

The aircraft industry is also having a run of disappointments after a period of uninterrupted progress, from the screw propeller aircraft in the first long cycle to the jet aircraft in the second. During the latter, the carrying capacity of planes has increased from a few score passengers to the jumbo jets (Boeing 747 or Galaxy). As a further increase in size is not possible, progress is being looked for in some other direction (as in the textile fibres industry).

Attempts have been made to achieve a major qualitative change, but without much immediate or prospective success. Apart from the setback with supersonic aircraft (Concorde and SST), the reasons for which are due to factors discussed in the next chapter, it is reported that the development programmes for the VTOL and STOL aircraft have been more or less abandoned by commercial aviation ([7], p. 153).

Progress in the motor car industry has not been as great as was expected in the 1960s. The Wankel engine has not been much of a success, and progress has been very slow in the projects for turbine engines and electrically powered cars. It is intriguing to compare the present position with

---

* "Chemical industry polishes its crystal ball", *New Scientist*, 3 February 1977, p. 276.

an earlier forecast based on past trends:

> "If forecasters of a decade ago had been correct, instant communications between cars and central computers, perhaps using satellite transmission, would be directing the flow of traffic on highways to ease congestion. Some cars already might have been equipped with automatic steering and braking and acceleration controls relying on radar-like electronic sensors to avoid crashes" ([7], p. 153).

Indeed, the present difficulties are all the more galling because of the expectations aroused earlier. The degree of disillusionment expressed in books such as the one by R. E. Miles [6] is only matched by the excess of optimism in the 1950s and early 1960s. H. Kahn [4] was then speculating on an era of automation, when disease would have been eliminated and the span of human life greatly extended. In one of his books, F. de Closets [1] includes a table of forecast dates for scientific and technical discoveries, prepared by the Rand Corporation in 1964. This gives 1971 for achieving desalinization of sea water, 1972 for automatic translation, 1973 for automatic air traffic control, 1974 for the generalized use of teaching machines, 1975 for reliable weather forecasting and aircraft propulsion by nuclear reactor, etc.

It is easy to laugh at such blunders after the event, but it must not be forgotten that, when they were made, they had a great effect on current thinking by painting the future in glowing colours. The gulf between the forecasts and the present reality of economic depression explains — though it does not justify — the reactions of those who blame capitalist profiteers or greedy union leaders for the failure to achieve a paradise.

Surely, the first question that all thinking people should ask is whether the conditions that made the technological progress of the last few decades possible still exist. This question may be approached in two different ways, i.e.

(a) Is technological progress slowing down?

(b) Is technological progress continuing, but becoming less and less efficient?

The two questions are not entirely separate because, if technology is becoming less efficient, this must eventually slow it down. All the same, we will examine them in turn, first in this chapter and then in the next.

To be able to say exactly how far there is a slowing-down in technical innovation, and the degree to which this is due to short-term and structural factors respectively (i.e. to estimate its likely duration), extensive research would be needed on the R & D situation in all sectors of industry. Without going as far as this, however, it is possible to form a fairly accurate idea by taking the sequence of events in the production of innovations (as outlined in the previous chapter) and studying the changes in the factors that govern it.

It can be seen from equations (9) to (10″) in Chapter 3 that a slow-down in technical innovation can only have three causes:

(1) A decline in expenditure on R & D.
(2) A fall in the research success rate.
(3) An increase in the time-lag between research and invention and between invention and innovation.

There is no longer any doubt about the existence of cause (1) above. Regular studies in the United States on R & D expenditures in industry have shown a marked and continuous slowing down in the growth of these over the years 1960 to 1975.

Period	Average growth rate of actual expenditure on R & D in the United States	
	Federal	Private
1950–60	13.9	7.7
1961–7	5.6	7.4
1967–75	3.0	1.8

Source: National Science Foundation and *Business Week*.

As might be expected, the averages given in the above table cover wide variations from one year to the next. Changing moods among management and cuts in research expenditure due to price rises can cause large short-term variations: for example, research budgets increased by 4.2 per cent in 1976 after being practically stationary in the previous year. But these variations cannot hide the clear trend in the figures in the above table, which can hardly be totally unrelated to the decline in economic growth in the mid-1970s. The phenomenon is not limited to the United States; it affects the whole Western world, though the main investment in R & D in

Europe took place around 1965 when the technological gap between the Old World and the New World was causing concern.

There are two areas in which we may find an explanation for the relative disinterest in research investment.

On the one hand, after heavy investment in the two previous decades, and faced with mounting liabilities connected with this, many companies are feeling a need to review the financial situation, especially as labour and raw material costs are rising.

On the other hand, certain developments in contemporary society have tended to cause a number of executives to change their attitude towards innovation. The growing interpenetration of politics and industry may be tending to increase the margin of uncertainty in judging business prospects. In addition, technical progress has been found to involve external diseconomies and social costs, and this does not encourage its pursuit at any price. These factors will be looked at more closely in the next chapter. For the moment it is enough to note that they exist and that they are inclining manufacturers to feel that research is not worth the effort — especially in cases where it is producing little result and the time-lag between research and innovation is tending to become longer.

As regards the second of the three possible causes mentioned earlier (a fall in the research success rate), it is difficult to provide any satisfactory pointers in the absence of research on this specific point. All that can be said is that, in view of the element of chance involved in scientific progress and in the chain-reactions from any fundamental discovery, this possibility cannot be excluded *a priori*. If the present period happens to be a "trough" in the progress of science, the effect of this will certainly be intensified if research in one sector can no longer benefit from the side-effects of research in other sectors. In equation (9) in Chapter 3, this hypothesis would appear as a fall in the ΔS factor and a negative bias in the random term u, with a consequent reduction in the flow of inventions or at least a slower increase in this flow.

It may seem paradoxical to think of a possible decline in the growth of the stock of scientific knowledge at a time when the number of research workers is increasing and when the general level of education has never been higher. A report of the Commission on Human Resources and Advanced Education published in the United States by the Russell Sage Foundation* drew attention to the explosion in post-graduate

* Cf. O. Giarini [2].

education, which leads to the Ph.D. degree and provides the bulk of recruits to university teaching and research posts. In 1920 only a few hundred Ph.D. degrees were awarded each year in the United States; in 1930 the figure had risen to about 2000, in 1945 to about 3500 and in 1950 to just over 6000. By 1960 the number exceeded 10,000, it was nearly 30,000 in 1970, and was close to 50,000 in 1975. How could there be a fall-off in scientific discoveries in spite of this?

The first point to be noted is that the "output" of doctorates does not necessarily correspond to the number of scientists employed. It measures the annual increase in the potential supply, but not every new holder of a Ph.D. wants to go into (or, even more, to stay in) the research field, and not all who wish to do this find a position that suits them. We know that for some years there has been a problem of graduate underemployment in the West, part of the reason being that the demand for research workers has not kept up with the rapid growth in the supply.

It is also common knowledge that apart from the increased output of degree-holders, there has been an increase in the average duration of university studies. This is certainly due to a number of factors, but it can mainly be attributed to three causes. Firstly, the expansion of educational opportunity could only be achieved as a result of a large number of students combining study with work in a job. Secondly, the development of mass higher education has encouraged a growing number of students wishing to postpone their arrival on a crowded labour market to continue their studies and improve their qualifications. Thirdly, the increase in the total stock of scientific knowledge is constantly making for longer periods of study.

Now, it has been proved that scientists as a rule make their main discoveries when they are young, and then work on them for the rest of their lives. Statistics of the ages of Nobel Prize winners when they made their discoveries show that, unlike the conventional picture of the scientist as a white-bearded elderly gentleman, they have nearly always been under thirty. Einstein published his first paper on relativity when he was a young employee of the Patents Office in Berne. Watson and Cricks discovered the structure of ADN at an age when less gifted students are finishing at the university. And the same can be said of the vast majority of Nobel Prize winners in chemistry, physics and mathematics.* Inventiveness depends

* Even among the Nobel Prize winners in economics, one finds that Samuelson, Leontieff, Hicks and Arrow, among others, wrote their fundamental books at the age of about thirty.

very much on biological factors. The mind of an individual begins to age biologically when he is twenty and he is normally capable of making inventions or discovering new relationships between scientifically definable and quantifiable parameters only up to the age of thirty. An invention in science or technology requires an unusual capacity for abstract thinking that depends very little on experience of human problems. This capacity may enable a scientist later in life to make use of his intellectual abilities in the humanities, but the golden age for inventions comes at a time of life that is too early in terms of the general social pattern.

The above points are borne out both by actual experience and by statistics. For example, the number of patents taken out by the largest laboratories depends both on size and on the age-structure of their research staffs. It can be calculated from statistical data than when the average age in an industrial and scientific research organization at the national level or in a major research institute reaches thirty to thirty-five years, its output of inventions (e.g. the number of patents as a ratio of the number of researchers) becomes extremely low. Beyond this age, the organization will have difficulty in maintaining its position in the field of active research. This general rule is not affected by exceptions or special cases.

Thus, the trend in the educational system is in conflict with the goal of higher productivity in research.

Moreover, even if there is no actual fall in the research success rate, the operation of certain economic and social factors to be discussed in the next chapter can lead in practice to the same result if they discourage the disclosure and/or utilization of any discoveries that may be made. Hence, the productivity of research can decline, even when a society's capacity for invention is undiminished.

Let us now turn to the third possible cause. When considering the eventuality of an increase in the time-lag between a discovery in science and its industrial application, we shall have to go back for a moment to the inter-relationship between research and development, bearing in mind that there are cases where technology continues to develop independently of science.

There are major differences between fundamental research, applied research and R & D, even if the boundaries between them are not always clear. Fundamental research is aimed at expanding scientific knowledge and increasing our understanding of Nature. Applied research then tries to put the discoveries of fundamental research to practical use, or even to acquire new knowledge likely to be of practical use. R & D enables the product of research to be exploited on an industrial scale.

In the United States, a little less than 10 per cent of all research investment in the past decade has been for fundamental research, while applied research has received about 25 per cent, all the remainder going to R & D. However, the distribution varies considerably from case to case, and the figures given indicate only the general tendency.

Sometimes fundamental research receives a stimulus from applied research (as in the case of fuel batteries) or even from R & D. Some writers accordingly make a distinction between pure research (without any practical motivation) and oriented fundamental research.

In addition, the final stage of development has to be linked up with market knowledge and the engineer must for the time being become an industrial economist. This is the market research stage, which is followed by marketing studies before the product can reach the consumer.

These distinctions are important for the design of a research policy or programme, whether in a firm or at the national level, in that they focus attention on the lead-time between the date when a research project is decided upon and the date when (if all goes well) it will produce results in terms of business.

The success due to technology over the past 200 years and the rising tide of innovations that have transformed the human environment have led to the idea that an acceleration principle is at work. The argument is often supported by a graph like the one in Fig. 2, showing the shortening of the time-lag between discoveries or new processes and their practical application.

The graph emphasizes and seems to explain the "shock effect" of technological acceleration and the reduction in the interval between research and innovation. Other studies have taken the same line. Gilfillan, a sociologist, showed in 1935 that the interval between the initial idea and its commercial success had been 37 years before 1900 and 9½ years after 1900. However, it was later noticed that the apparent acceleration was due to the incompleteness of the data used, and other studies threw doubt on Gilfillan's argument. After studying thirty-five inventions, John Enos found, for example, that the average time-lag before 1900 had been

12½years for twelve inventions; between 1911 and 1940, it had been 15.9 years for eleven inventions; and for twelve inventions covered since 1941 it had fallen again to 12½ years. Thus, no conclusion could be drawn as to a possible trend in the time-lags.

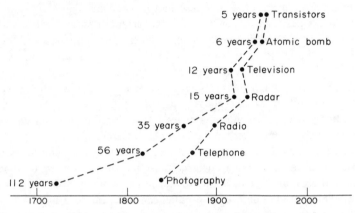

Fig. 2. Examples of the interval between the discovery of a principle and its commercial development.

Source: L. Hafstad, "The role of industrial research", *Science Journal (London)*, September 1966.

In an article on this question, J. P. Martino [5] showed that the inconsistencies between the results of different studies are attributable to sampling defects and the difficulty of defining an invention. It points out that the above graph is meaningless since the choice of starting-dates is quite arbitrary. The year 1725 was chosen for photography because that was when it was first discovered that silver salts turn black under the influence of light and not — as previously thought — through the influence of heat. But the beginnings of photography could equally well have been set back to the discovery of silver salts in 1565 or to the invention of the camera obscura in the year 1000, or alternatively set at the date of the first photograph in 1822. Similarly, the atomic bomb was dated from the Manhattan Project in 1942, but there had been a series of discoveries since the beginning of the century which had foreshadowed the invention. As regards the choice of sample, Martino points out that:

"Even if there is no trend whatsoever in the rate of adoption of inventions, a study which only looks at average time lags for different periods is almost bound to find an increase in the rate of adoption for the more recent periods, because of the inherent bias arising from omission of inventions whose adoption has not yet occurred" ([5], p. 71).

All this leads to the conclusion that there has not been up to now any irrefutable proof of a reduction of the time-lag between research and invention. Our feeling of accelerating technological change comes from the interaction of two separate phenomena, one in the cultural background and the other in the technical field proper. On the one hand, both manufacturers and the public have become quicker to adopt new inventions and to make use of them in daily life. In the technical field, on the other hand, the opposite appears to be taking place.

Until quite recently, the speed of change in attitudes was more marked than the technological slow-down. Firms became increasingly responsive to technical progress, and followed up all usable inventions in the public domain. They could draw on both the flow of new inventions and the stock of earlier discoveries neglected by their predecessors. This created an illusion of a shrinking of the time-lag between research and innovation.

Today, the cultural time-lag is practically nil, so that the trend in technical lead-times alone is apparent — which is the aspect with which we are concerned here. It is becoming clear that the time needed to develop a new technology increases with its dependence on scientific knowledge. Physicists have confirmed that any discovery in the field of fundamental research is unlikely at the present day to become usable in less than twenty years' time. Moreover, discoveries are rarely the result of sudden inspiration. They are mostly the outcome of a long search taking decades and progressing step by step with occasional leaps ahead. Any prediction as to when a discovery will be made is, as noted earlier, almost entirely arbitrary. And the example of the argument as to whether the main credit for the discovery of nuclear energy should go to Madame Curie, or Niels Bohr or Oppenheimer or Fermi, shows how complex the background to a discovery may be.

Even when a new technology has reached the prototype stage, it will often be several more years before the public sees the manufactured product. If we take the new generation of "shuttleless" looms, for example, the first ideas for the type that has been most successful appeared over a period of about ten years (1930–40). The first prototypes were developed between 1945 and 1960. Sales of the first machines began in 1960, but it was not until 1975 that these reached the figure of 10 per cent of the world market. An innovation that is still felt as a "future shock" by some traditional textile manufacturers was already on the horizon forty years ago. Similarly for many people, it comes as a great surprise to be told that television was ready for use before the Second World War.

Thus, for reasons of social psychology connected with the way in which industrial information is disseminated, changes that were not unexpected have come upon the public without warning. In reality, progress in technology is now much more difficult and takes much longer to achieve than in 1900. As the successive stages between fundamental research and the final user increase in number and complexity, a multi-disciplinary approach becomes increasingly necessary for new technologies, and research costs become higher and lead-times longer, even though industry and the public have in the meantime made enormous psychological progress in foreseeing the possible uses of a new invention. Consequently, a phenomenon that has been depicted as an acceleration of scientific and technological progress is to a great extent a reflection of the quicker perception by business leaders of opportunities involving the use of *existing* technology.

Seen in this light, the graph showing an accelerating reduction in the time-lag between research and innovation is misleading. It hides the inevitable increase in the time-lag when further progress in technology demands increasingly complex scientific research. We should not forget that the Apollo programme for a landing on the Moon could only be achieved in 1969 by using technologies that already existed ten years before, since the fundamental research carried out during the ten-year period could not be applied in time.

It can, of course, be argued that an increase in the time-lag need not slow down technological innovation in so far as investment in R & D goes on increasing and the productivity of research can be maintained. The crux of the matter is precisely that (as we have already seen) the growth of investment in R & D is fluctuating, and it is not certain that the productivity of research can be maintained.

Hence, the increase in the time-lag between research and innovation is a decisive factor in the analysis of diminishing returns to technology, and we shall now see that it is not without effects at the micro- and macro-economic levels.

Reverting to the traditional analysis of investment in research, we find that there is a correlation between two series of magnitudes, i.e.

the percentage of total sales of an industry invested in R & D;

the rate of obsolescence that this investment causes in this industry.

The following table, based on estimates and returns in Britain in 1965 illustrates this relationship.

Industry	Percentage of total sales invested in R & D	Product obsolescence (years)
Research	100	0.1
Aviation	40– 50	1
Electronics	15– 20	5–10
Electrical engineering	10	19
Chemicals	5	60
Non-ferrous metals	2	100
Food	0.6	100
Textiles	0.3	100
Paper	0.25	100

We can see from the table that

(a) the cost of R & D in certain industries exceeds that of the other factors of production;

(b) the obsolescence of the product (in years) is inversely proportional to the amount invested in research.

The table can be shown as a graph in which the different sectors are ranked on the basis of the two parameters (Fig. 3).

Looking at the graph, one begins to see the nature of the problems encountered by the most advanced industries in recent years. Any industry that moves leftwards along the curve must sooner or later reach a point where the system of investment in R & D becomes self-defeating. Let us try to see this more clearly with the help of Fig. 4.

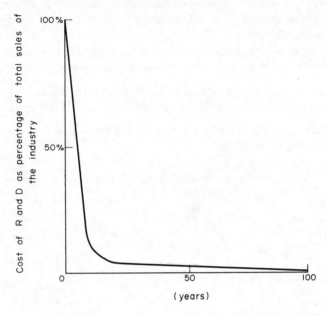

Fig. 3. Average obsolescence period of the products of the industry.

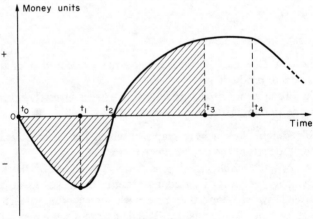

Figure 4

- At point t_0 a research programme is begun. The money is invested in it but no production (and so, no earnings) can be expected before time t_1. This period is generally several years in length.
- At t_1 the research programme has started to show commercial results, but research is continued in order to improve the reliability of the product, so the firm continues to spend more than it earns and its losses increase.
- At point t_2 the firm's current expenditure on research, production and distribution is offset by the proceeds from sales.
- The accumulated deficit (the shaded area below the line t_0 to t_2 on the time axis) is not made good until t_3. The firm now begins to make a profit. But when the initial decision was made to begin a research programme, this point lay far ahead in the future. For research on a considerable scale (such as for the Concorde) the interval between t_0 and t_2 may be ten years or more.
- At point t_4 a new product comes out, making the first product obsolescent and eliminating it from the market.

This is, of course, an over-simplified picture, but it gives an indication of a number of trends mentioned earlier that are increasing the vulnerability of the industries using advanced technology.

In the first place, the research programme starting at t_0 must make allowance for the fact that, on average, less than 10 per cent of the research will be successful. A manufacturing company that invests $100 million in R & D expects $90 million to be a sheer loss. The part of the programme that is commercially successful must therefore recover, in addition to its own cost, the cost of all the failures. If the percentage of success declines, the position becomes more serious since the company will need to increase the size of the research programme continuously to take advantage of the law of large numbers. Otherwise, with research confined to a few areas only, results become more or less a matter of luck. Most firms are in fact obliged to accept the second solution. Certain electronics companies, for example, eventually decided that their research programmes would in no case represent more than 25 per cent of their turnover. In the alternative case of a company wishing to maintain the same flow of innovations, the position quickly becomes untenable: since obsolescence does not vary, point t_4 is fixed; on the other hand, the area of financial loss in the graph continues to grow as the research programme is expanded, and research finally ceases to be worthwhile.

In the second place, even if the success rate in research does not fall, allowance has still to be made for the gradual lengthening of the time-lag between research and innovation. The effect of this is to move t_1 in the graph towards the right on the time axis, so increasing the area of financial loss. If the obsolescence period gets longer at the same time, the problem is less serious, as the manufacturer will have more time to recoup the increased research costs. But he can never be certain that a competitor will not have a stroke of luck and complete an R & D programme earlier than expected.

Any lengthening of the development period for a new product or process therefore increases the manufacturer's uncertainty as to whether his R & D programme is worthwhile. Apart from this factor, he must also consider other uncertainties: technological progress among competitors and among manufacturers of substitutes, and possible changes in market conditions (consumer attitudes, action by the government or pressure groups).

Hence, when a research programme proves successful, it cannot be assumed that the consumer will benefit from the savings in production costs due to the new product or process. These must enable the firm to make up for deficits on other past or future programmes. The introduction of a new technology no longer implies automatically a reduction in price. This helps to explain findings that the relative prices of manufactured goods, after declining consistently in recent decades, had been tending to rise for about five or six years.*

This brings us face to face with one of the inconsistencies of an industrial system dependent on the discovery of an increasing number of new technologies. Sooner or later, a limit is reached and disappointed manufacturers turn away from investment in R & D.

A further consequence of the lengthening of the time-lag between research and innovation arises at the macro-economic level.

Economic theory tells us that when a good becomes scarce on the

---

* A further point worth noting is that the increasing interval between research and innovation lengthens the period during which purchasing power is being made available without any corresponding addition to the goods or services produced, so that its effect may be inflationary.

market, its price rises, so discouraging surplus demand and stimulating supply. On the face of it, this mechanism should encourage substitute products and the introduction of processes making it possible to increase the supply of the scarce good.* The problem is that this model is essentially a static model, making no allowance for time-lags in the response of supply to demand. The great English economist, Alfred Marshall noted this and introduced the distinction between the "short- term", when the curve of supply is inelastic, and the "long term", when it becomes elastic.

Now, if the time-lag inherent in the R & D process increases, the economy becomes more inflexible. If a block occurs anywhere in the system, its elimination by means of a new technology will take longer if the new technology takes a long time to perfect. In the case of an unforeseeable (or at least unexpected) rise in oil prices, for instance, a number of years must elapse before nuclear power can be substituted. When it is then found that nuclear reactors are regarded with hostility or scepticism by a large part of the population and many scientists consider them dangerous, it turns out that a change-over to solar energy will need ten or twenty years of research.

We are therefore faced with a significant fact, i.e. that industrial development based on a scientific type of technology leads in the long run to a loss of flexibility in the economic system. It becomes increasingly difficult for the system to react spontaneously to an emergency because the market mechanism is held back by technological delays. And the increased vulnerability of the economic system makes it increasingly necessary to resort to planning, both at the company level and at the national level.

When reflecting on the mysterious nature of time, one of the questions that St. Augustine asked himself was: "If the future and the past are real, where do they exist?" The increasing significance of the time factor in economic life makes his question more relevant now than it was in his own time.

If we follow up this line of thought by asking what factors set the pattern for the present at any particular time, we could say that it is the result of human actions or inaction and of chance events — according to the relative importance of each and how long its effects persist. Thus, mankind is in a sense continuously "planning" the future situation, unconsciously or consciously. If a manufacturer decides to set up a motor car factory in a

* If the factors of production are already being fully employed.

particular city, he is in effect predetermining part of the future of the city.

The point behind these remarks is that the main problem in planning is that of evaluating the "inertial" or on-going effect of present actions and events. A planning strategy will fail if it does not estimate the different lead-times and identify the time zones when these will make way for new decisions (Fig. 5).

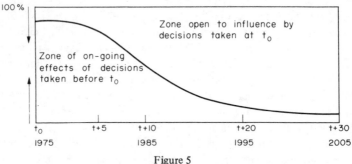

Figure 5

The reason why so many planning exercises fail is that they are based on a series of objectives, without regard to the factors in the present situation which have already set the pattern for a specific future, the time during which these will be operating and the probability of unforeseeable events.

Lead-times can, of course, be modified after they have begun to run, but their degree of flexibility has to be considered. At the level of the global society, ignorance on this point is the main cause of political excesses, whether by revolutionary dictators seeking to eliminate hang-overs from the past (e.g. Robespierre and Stalin) or by conservative dictators seeking to perpetuate the past (e.g. the last Tsars and Franco). In the first case, the limits of flexibility are exceeded, and in the second case the aim is to eliminate flexibility. In both, the ideology becomes a screen for rudimentary attitudes.

The connection between technology and what we have been discussing is that, when an organization or an economy reaches a stage where there are increasing problems of timing as a result of science-based technology, a new approach to planning is needed. It has to be realized that every activity and event sets a pattern, and that the only kind of forecasting that can indicate the range of possible futures is one that allows for this fact. The range becomes narrower in the short term as the increasing time-lag between research and innovation reduces the ability of technology to affect the future. Apart from being largely responsible for changing the pattern of society and the operating conditions in the economy, technology has finally affected the most characteristic feature of the society, i.e. the pace of change.

In concluding this chapter, we should stress the fact that there are two phenomena which may be having a retarding effect on technological innovation, namely, the decline in investment in R & D, and the increasing lead-time needed for developing innovations. The latter is of particular importance as it is a logical consequence of developments in science-based technology.

Only quantitative research could tell us whether we are facing an actual decline in innovation or simply a decline in its rate of growth. It is worth noting, however, that the second alternative by itself would to some extent explain the disordered state of the societies which had based their objectives on the likelihood of constantly accelerating technical progress.

A slowing down in innovation is not yet obvious to an observer of the industrial scene, but this may be due to a time-lag in our ability to interpret new situations that do not fit into the customary framework of analysis.

In any case, the two phenomena operating in this direction are not the only reason for anxiety. There is also the overall benefit from technical progress. If this declines as a result of a combination of factors, there is reason to fear a general loss of interest in R & D, leading automatically to a gradual slow-down in innovation.

References

1. François de Closets. *En Danger de Progrès.* Gallimard, Paris, 1972.
2. Orio Giarini. "Mythes et réalités du gap technologique", *L'Europe en Formation,* Paris, July—August 1971, pp. 20—25.
3. Stanley Hurst. "Beware of the general-purpose-micro-processor", *New Scientist*, 10 February 1977, pp. 322—324.
4. Herman Kahn and Anthony J. Wiener. *The Year 2000.* McMillan, New York, 1967.

5. Joseph P. Martino. "The pace of technological change", *The Futurist,* April 1972, pp. 70–72.
6. Rufus E. Miles. *Awakening from the American Dream. The Political and Social Limits to Growth.* Universe Books, New York, 1976.
7. Donald Moffitt (Ed.). *America Tomorrow.* Amacom, New York, 1977.

CHAPTER 5

The Decline in the Over-all
Efficiency of Technology and the
Problem of "Value Deducted"

5.1. The limits of technical efficiency

Technical progress operates in three main directions:

1. It enables new products to be marketed, thus increasing the range and quality of goods and services available to the consumer. The introduction of television, for example, was a net addition to the sum of goods and services available, since it did not lead to the disappearance of any of the facilities which it to some extent replaced, i.e. radio, theatres, cinemas, newspapers, etc. In the same category is the marketing of washing machines with automatic spin-drying, which represented an increase of utility for the consumer as compared with the earlier machines with a hand-wringer.

2. It provides new processes for the production of goods or services, in which human labour or animal power is replaced by machines, or which improve the productivity of existing machine operations. Processes of the first type led the way to the Industrial Revolution in the middle of the eighteenth century and have steadily increased up to the present day in the three main branches of production: agriculture, manufacturing, and services. Their effect has been reinforced by the second type, which covers a vast range of innovations – from the Thomas-Gilchrist furnace to the high-capacity computers. Scientific work management methods also belong to this category since, by de-individualizing the tasks, the workers can more easily be integrated in a production line yielding high labour productivity.

3. It provides ways of making available to the economy fresh sources of energy and raw materials. Examples are the production of

electricity through harnessing nuclear energy and recent developments in the exploitation of ocean resources: improved fishing
techniques, boring for off-shore oil, desalinization processes, dredging
for ores on the sea-bed, etc. [18].

Technical progress is a product of human activity and, as such, gives rise
to externalities in varying degrees, in the same way as the working of a
quarry or ability to play the violin. Externalities can be broadly defined
as the side-effects of one activity on other activities now or in the future.
The effect may be positive (external economies) or negative (external
diseconomies). The British economist, J. Meade [28], has provided an
idyllic illustration of the former: external economies are created if a
horticulturist plants his flowers near the hives of a bee-keeper; the bees
will have ample supplies of pollen near at hand and there will be more
and better honey. As an example of external diseconomies, economic
textbooks tend to quote the example of a laundry where the work suffers
as a result of a nearby factory chimney spreading black smoke over the
area.

External diseconomies do not necessarily imply direct financial loss.
They also exist in the more general form of nuisances. These have been
attracting more public attention in recent years, so that there is a tendency to associate technical progress with the creation of external diseconomies, such as environmental pollution and annoying side-effects
of new products (such as noise). Without minimizing these, one must
add that technology also creates external economies. In the business
world, for example, the use of computers has led to expanded sales of
air conditioning equipment. In the social field, progress in medicine has
raised the educational level by reducing the incidence of disease on the
mental development of the young. And from the environmental point
of view, higher productivity has reduced the number of factories.

Since technical progress has both direct and indirect effects, these
must all be taken into account when judging its efficiency at a given
moment and the trend in its efficiency over time. There are a good many
indications today that the beneficial direct effects are becoming less
marked, and that the external diseconomies are increasing.

Let us first take a look at the good effects of the three main forms
of technical progress indicated earlier.

In so far as it is still productive, modern technology is still providing new products which increase utility for the consumer, but the additions to utility seem to be following the law of diminishing marginal utility familiar to economists. The advent of the radio must have been a sort of "future shock" for our grand-parents since it immeasurably increased their opportunities to know about the wider world and its events. In comparison with radio, television also represented a major advance, but only an extension to the visual field of the access to the outside world already provided by the radio. When colour television arrived, this marked further progress but no one would regard the change from black-and-white as comparable with the change several decades before when the radio set was installed in the kitchen or sitting-room. The same certainly applies to the introduction of faster, more comfortable and more advanced automobiles, and to the change-over from manual to automatic telephones (and perhaps soon to the videophone). Hal Hellman [25] made a similar comment in connection with faster means of transportation:

> "Thus we can say that the American SST would have flown three times as fast as a modern subsonic jet, which sounds like an extraordinary, shocking change. But the *savings* — for the élite few who would have used it — would have been four hours. Now put yourself back in the nineteenth century. Crossing the country by horse and wagon took an average of six months, by ship around South America it took three months. But in 1869, when the last spike was driven to complete the first transcontinental rail line, the time was cut to six days. The point, of course, is that not only was the railroad thirty times faster than the previous land method, but the saving was more than five and a half months" ([25], p. 10).

These examples suggest that, even if technology continues to produce results, the perception of progress is necessarily declining in a society where, contrary to the view expressed by Alvin Toffler [36], "future shock" is now a thing of the past.

Looking now at the type of innovations that expand the stock of usable natural resources, one finds the same phenomenon occurring and a consequent tendency to minimize the benefits from this type of technical progress. We have already mentioned drilling at sea and nuclear energy as examples. Now, public opinion does not perceive these as genuine improvements but as part of the maintenance costs of the economic system: oil must be got from the ocean because oil from the land areas is coming to an end and the reserves are in the hands of the Arab countries, and nuclear power must be developed to keep up with electricity requirements and also — it is said — to lighten the burden on the balance of trade. Technology is not regarded as a means of increasing utility but as the only way of maintaining what is regarded as the "normal" upward movement of living standards.

Thus, in two key sectors of technological innovation, the community no longer attaches so much value to the achievements of technical progress: its faith in this is declining and it is less conscious of the direct benefits that it can offer. It should be noted that this is taking place quite independently of the indications of a slow-down in technological innovation discussed in the previous chapter.

But — you will say — what about the continuing direct effects of technology in the production of goods and services? It cannot be argued that the productivity increases here are of diminishing marginal utility.

All the same, we must note in the first place, that the adoption of new, more efficient processes leads at times directly away from the intended goal. We have shown in another context [18] how progress in the area of renewable natural resources can lead to over-utilization and eventually to declining production. Modern fishing techniques are so efficient at locating and catching fish that fish stocks have been greatly reduced (off the coast of Peru, for example) and catches have fallen since the peak in 1970. The formation of deserts in various parts of the world is also due in part to over-working of the soil as a result of modern methods of cultivation.

Indirectly, innovation in production techniques may also be more than off-set by the structural changes that it leads to in the production system.

The implicit goal of the industrial revolutions in the eighteenth and nineteenth centuries and of all economic development since then has been to eliminate scarcity by producing the maximum possible amount of consumer goods. Adam Smith himself pointed out in 1776 that the aim of economic activity is consumption. Saving is not a refusal to consume but a postponement of consumption: savings both by individuals and the community lead to investment, so enabling consumption to be maintained or expanded. Hence, the problem in each economy has traditionally been stated in terms of maximum production with the scarce resources available. The justification for competition was held to be that, if entrepreneurs seek to maximize profit, this enables a larger quantity of goods to be offered, at lower prices. This view naturally led to physical production being regarded as the principal form of economic activity, and the two successive industrial revolutions to be valued primarily for having enabled more to be produced at lower cost. New processes allowed — and to some extent still allow — more to be produced per unit of time, with the same amount of labour and capital.

The problem is that technology has been so successful in this aim that it cannot go much further. The limits are reached when physical production becomes so efficient that it ceases to be the dominant sector and becomes a marginal sector of the economic system. Much fewer resources in labour and capital are now needed for physical production but, as the output has become enormous, many more resources are needed for management, distribution and marketing. Now, most innovations in the field of processes still relate to the physical production sectors, and so — no matter how *technically* efficient they may be — affect a shrinking part of the whole economy: hence a progressive decline in the *return to the economy as a whole*.

The above tendency is reinforced by the fact that technical progress in many cases involves increasingly specialized processes. Every new solution tends to offer advantages in a narrower field of application than the technology it replaces. In the textile industry, for example, the traditional all-purpose loom has been replaced by a generation of shuttle-less looms that differ according to the product. The limit of specialization is reached when a single machine can produce 500,000 blankets a year — or the total consumption of a small country — as long as all the blankets are of the same material, pattern, size, etc.

And to come back to products again, Stanley Hurst describes how the manufacturers of micro-circuits were recently faced with a problem of extreme specialization:

> "Their expertise had advanced so successfully that single monolithic circuits containing thousands or even tens of thousands of active devices have become possible. But the

total cost of bringing such circuits to the market-place is in the million-dollar bracket. Therefore, to bring the unit cost of complex circuits down to a commercially viable level, a large demand must exist or be created for each individual product. But the larger the circuit, the more specialized it is likely to be — and the smaller the global demand. Hence the dilemma" ([26], p. 323).

In cases of this kind, specialization is still attractive from the production point of view, but it is an absurdity from the point of view of distribution and utilization. Production of an increasingly specialized range of goods more rapidly and on a larger scale increases raw material requirements both in quantity and quality at each stage of production; it calls for heavier investment in storage facilities for inputs and the outputs; it involves wider dispersion; it increases the number of intermediaries; and it aggravates the whole set of financial problems connected with the change. Hence, it is becoming increasingly rare for the cost of manufacturing a product to represent more than 20 per cent of the selling price to the user, the bulk being taken up by a variety of distribution costs.

"Why do chips cost so much more than raw potatoes? Because the final product embodies a number of services: research and development of methods of preparing and conserving the product, organization of a distribution network, and marketing activities" ([4], p. 41).

All this suggests that it is a mistake to apply cost-benefit analysis to a new technique solely at the production level. Savings of 20 per cent at the manufacturing stage where this stage represents only 20 per cent of the total cost are wiped out if distribution costs increase as a result of the change by as little as 5 per cent. The reasons which dissuaded the chemical industry some years back from continuing building plants with larger and larger capacity for the production of ethylene, ammonia and other intermediate goods were of much the same kind.

In short, the enormous success of the marriage between science and technology has opened the way to a vast expansion of service activities which completely overshadow the effects of innovations in manufacture. This trend is well-known from the books of Colin Clark [7], Fourastié

[12] and Daniel Bell [3].* It is confirmed by statistics: in 1969 the tertiary sector represented 57 per cent of the labour force in the United States and 42 per cent in France; in 1974 these percentages had risen to 62 and 49 per cent respectively. A point less often mentioned is that, quite apart from the growth in the tertiary sector itself, there is a growth of service activities ("tertiarization") within the manufacturing and agricultural sectors. The phenomenon is particularly marked in the Federal Republic of Germany, but it is taking place in all the industrialized countries – for example, in France for which some figures are given below:

Percentage of tertiary posts in certain non-service industries in France, 1971	per cent
Fishing, forestry, agriculture	28
Water, gas, electricity	39
Petroleum and liquid fuels	44
Mechanical engineering	17
Electrical engineering	17
Glass manufacture	16
Building and public works	11
Chemicals	30
Food	32
Textiles	13
Wood and furniture	14
Leather	11
Paper and paperboard	17
Printing	34
All industries covered	34

Source: INSEE.

* It is interesting to note that technical progress has been a major criterion in the division of the national economy into three major sectors. Fourastié, for example, writes:

"I shall use the term "primary sector" to cover activities of an agricultural type since this is a traditional sector and still an essential one for food and clothing; this is a sector with average technical progress judged over a long period. The "secondary sector" will cover the industries with a high rate of technical progress: it is therefore for practical purposes the manufacturing sector. Finally, the "tertiary sector" includes all other activities, i.e. those with little or no technical progress....The tertiary sector is actually very extensive; it covers commerce, the public service, education, the liberal professions and a great number of manual crafts" ([12], pp. 81–82).

It is clear from the above figures that the statistics of employment by major sectors give only a partial picture of the distribution of manpower between them. Whether a typist or an engineer is classified under the secondary or the tertiary sector depends on whether she or he is working in a factory or in a research firm.*

Worse still as Philippe Grosjean points out — there is increasing inter-penetration between the sectors:

> "The secondary sector includes an increasingly large percent-age of persons belonging to the tertiary sector, while owing to mechanization the latter increasingly needs personnel of the kind hitherto found only in the secondary sector" ([22], p. 20).

However, as the increase in secondary-type jobs in the tertiary sector is much less than the growth of tertiary-type activities in the primary and secondary sectors, the over-all effect is a "tertiarization" of the economy as a whole. The process is partly the result of increased productivity in physical production, as we have seen, but it is also due to a shift of de-mand by households towards the service industries after saturation with agricultural and manufactured goods. Since productivity in the tertiary sector has up to now been progressing at a much slower rate than has been usual in agriculture and manufacturing, there has been a slowing down of growth, as described by Jean Denizet:

> "Growth over the last hundred years — and especially the last 50 years — has responded to a particular pattern of household demand. With rising household incomes, the demand has been less and less for food products and more and more for manu-factured products, especially durables. Now, up to the Second World War, agriculture was typically a sector with rising costs. Manufacturing, on the other hand, was from the beginning a sector with falling costs (mainly because of economies of scale achieved without higher productivity per worker, but also through increased productivity due to better work organi-zation and equipment, etc.). Because of the major part played

---

* Without counting the fact that — in the iron and steel industry, for example — the majority of workers are employed on inspection and maintenance work (cf. J. Parent [31]).

by economies of scale due to mass production, the shift of demand from low-productivity agricultural products to high-productivity mass-produced products resulted in extremely large gains in productivity for the economy as a whole, and these were the main influence in the growth of the national product over the last five decades...The shift of household demand towards the service industries where productivity per unit is high but constant makes a continuation of this model of growth impossible" ([10], pp. 88–89).

Denizet is here referring to only one of the two aspects of tertiarization but the gap is filled by Octave Gelinier [13] when he points out that productivity in agriculture and manufacturing are increasingly dependent on their tertiary component.

The trend towards diminishing marginal productivity in the economy as a whole might certainly be checked if a series of innovations occurred in the tertiary field. Electronics has already enabled a major advance to be made in this direction, and some think that productivity in the services sector is bound to rise substantially in the future as a result of large-scale investment in rationalization and modern management methods [4]. However, this will not be easy to achieve, since R & D has up to now been oriented toward new products and processes – quite understandably since the second industrial revolution mainly took advantage of advances in the natural sciences. Among the service industries, only transportation has directly benefited from the marriage of science and technology. The development of electronics has helped both manufacturers (automation) and services (data-processing) but this again is a fall-out from progress in physics and chemistry. This makes one wonder whether the future increase of productivity in the tertiary sector will not be bound up with advances in the human and social sciences.

However that may be, the fact that the high-productivity sector is forming a smaller part and the low-productivity sector a larger part of the economy as a whole is only one aspect of the question. The present limits to the efficiency of technology in physical production are also due to the fact that the logic of increasingly large-scale, high-speed, concentrated and specialized production is causing difficulties up-stream and down-stream from actual production. Innovations that could further improve performance turn out to involve disproportionate drawbacks at both the micro- and macro-economic levels.* The theory of economies of scale on which the process is based operates only up to a certain point. Thereafter, the advantages from the production angle are more than equalled by the disadvantages in the area of warehousing, distribution and the economic environment in general. Thus, a point is reached when the side-effects of technical progress have to be considered. It is then found that, to restore the over-all efficiency of technical progress, not only must the productivity of the tertiary sector be increased but production technology must be reoriented.

The most obvious aspect of the problem of the externalities of technology is damage to the environment, i.e. pollution. Until recently, this side-effect went more or less unnoticed because the planet's ability to absorb pollution was sufficient when the population was comparatively small and per capita consumption was low. The more rapid rate of population increase during the last few decades has suddenly brought us face to face with the imminent destruction of our natural environment, and the risk of an irreversible disturbance of the ecological balance.**

Technical progress has played a part in this because it has enabled a

---

* Owing to the fact that our system involves specialized and therefore mutually dependent units, such innovations may unfortunately be introduced: "It is widely assumed that improving the performance of part of an economy will lead to an improvement in the performance of the whole. The fallacy here is that, within a complex system, optimization of a subsystem is almost certain to depart from optimization of the total system" (Peter Chapman [6], p. 258).

** Garett Hardin [24] has shown that the Western economic ethic has played a part in the degradation of the environment and the depletion of natural resources. The "invisible hand" doctrine of Adam Smith and all the Classical Economists is that every individual, by pursuing his own advantage, promotes the public interest. Things appear in quite a different light if the concept of the economy as a closed circular flow is abandoned and the natural resources of the environment are included in the picture. A series of decisions that are rational from the individual's point of view may in fact be squandering of the world's resources and a loss in over-all welfare (see also Giarini, Loubergé and Schwamm [18], Chapter 4). In Solow's view [34], the problem of natural resources will be solved by the price mechanism encouraging substitution. But, as Georgescu-Roegen [17] points out, this overlooks the fact that any type of production requires natural resources of some kind or other. In other words, the depletion of natural resources will lead in the long run to higher relative prices for goods as compared with non-material services. If so, the outlook for domestic work, which has been so badly affected by the Industrial Revolution, might be very promising.

larger population to live better. But science-based technology has further complicated the problem of pollution by developing materials with a new structure that is resistant to the normal process of natural recycling.

The production sequence, from the extraction of the raw material to the utilization of the final product, involves a series of transformations which produce waste as a by-product (see Fig. 6). In the case of a raw material, this occurs even at the extraction stage, and it recurs at each intermediate stage. Sometimes the amount of waste is greater than the usable product. Eventually, when the final product has been used, it also becomes waste in one form or another. The fact that this problem is getting worse is, incidentally, why the chemical industry is devoting so much attention to possibilities of re-using waste by reprocessing it into by-products. Even so, the re-processing operation and the by-products themselves will inevitably produce waste.*

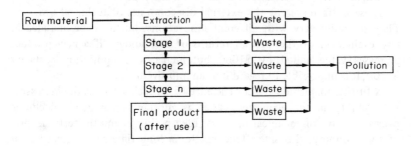

Figure 6

Now, the problem of delays or time-lags associated with science-based technology turns up again in connection with waste. The environment has a certain capacity to absorb pollution, but this process takes time. What is significant here is that not only is the volume of waste increasing but the time needed for natural recycling is getting longer: where scientific technology changes the chemical structure of natural materials, the product often takes longer to break down, e.g. plastic compared with wood. This problem is greatest in the most advanced area of technology, i.e.

---

* Waste is used here in the widest sense, i.e. solid or liquid waste, waste gases and waste heat.

nuclear energy, where the destruction of atomic waste takes centuries. In addition, the concentration of production in large units to take advantage of economies of scale has made pollution more severe in the areas affected and the natural recycling process longer.

Thus pollution is the result both of quantitative and of qualitative changes brought about by technology. It is now on such a scale that it cannot be disregarded as in earlier decades, and is now a major issue in politics and industry. It has become part of the costs of the production system and not merely an incidental "nuisance". A Battelle Memorial Institute study showed that pollution control by United States industry cost $4 billion in 1975 and was expected to reach $17 billion (at constant prices) by 1985. About 90 per cent of the total is the cost of implementing ten anti-pollution standards adopted. In cases where the cost is not borne by the industry concerned, it is to be covered by taxation, as happened after the wrecking of a tanker off the coast.

These costs represent the external diseconomies of technical progress. They naturally vary with different innovations, but must be included in any evaluation of the over-all returns of technology. If a new product increases value added by $1000 but increases anti-pollution costs by $1200, it is impossible to regard it as an improvement.

A further major side-effect of science-based technology is the increasing vulnerability in the economy, which is another factor in the declining over-all returns of technical progress. Like pollution and the tertiarization of the economy, this side-effect was for a long time so marginal that it could not affect the picture of technology as the main source of increased welfare. In the course of time, however, as the industrial revolution penetrated more deeply into society and the logic of concentration and specialization proceeded, the defects of the system became more obvious and are now beginning to outweigh its advantages.

The term "vulnerability" is used here to describe the situation of a "socio-economic system" (a firm, a group of companies or a national economy) where its operations or even its survival may at any moment be jeopardized by a chance occurrence due to a human cause (error, sabotage, etc.) or to a natural catastrophe (cyclone, earthquake, etc.). A system's vulnerability to a hazard has nothing to do with the probability of the hazard occurring. Even where this is highly improbable, the system is vulnerable if it might in that event be destroyed or seriously crippled for a long time. Vulnerability implies a rise in the level of uncertainty and a drop in the level of reliability. It does not depend on the frequency of a loss-causing event [21]. Situation (A) where the probability of winning $10 is 0.10 and of losing $10 is 0.90 is less risky than situation (B) where the outcome is either a win of $20 with a probability of 0.90 or a loss of $1000 with a probability of 0.10. The probability of winning in situation (B) is much higher, but the standard deviation in the outcome is $306 as compared with $6 in situation (A).

Insurance techniques take this fact fully into account. The most easily insured risks are not the ones that rarely occur. The latter generally involve large amounts in a market that is very limited (e.g. super-tankers) so that the law of large numbers does not operate and insurers have to add a large safety margin to the premium. Insurance companies prefer to cover highly probable risks such as car accidents, for which reliable statistics are available and the small amounts involved (in relation to the insurer's ability to pay) make it possible to spread the individual risks over the insured community.

Here again, nuclear power is a good illustration of what we are discussing, since it represents the most drastic case up to now. Even allowing for the strong feelings provoked by this thorny question, when a writer like Beckmann [2] praises the "safety" of nuclear reactors because an accident is theoretically less probable than when other sources of energy are used, he is simply playing on words. Even if the 'mathematical expectation of loss in terms of human lives is less high than with oil or coal — and this has still to be verified *ex post* since the Boeing 747 was the "safest" aircraft until the Tenerife catastrophe — nuclear power stations nevertheless represent increased vulnerability for the following reasons: (a) the maximum possible loss is greater (it could paralyse the life of a whole region); (b) accidents are rarer and so there is less opportunity of gaining experience in dealing with them — the only possibility is to imagine the scenario of possible accidents and base simulated alerts on it; (c) the production of nuclear power is concentrated in a small number of large-capacity units, so increasing the variability of the supply of electricity in a world where the part played by imponderables is still very large; (d) the techniques are so complex that expert knowledge is needed to understand them and the public has little say on a development that is likely to become an indispensable part of its way of living; (e) the construction stage is long and costly but the reactor has only a short life; (f) the pollution potential is greater because of the time needed for the waste to decay and the amount of waste heat generated; (g) the danger potential is a temptation to terrorists; (h) the technology used and the scale of production make it impossible to turn back, once the process has started, in spite of the fact that some technical problems are still unsolved and must be left for future scientists; (i) the controversy aroused is producing a sharp divergence between the direction in which the economy is moving and the feelings of a considerable part of the population who form part of the system and for whom it functions. This is not the place to present the case against nuclear reactors; their relevance here is as an example of the growing vulnerability. It is (or should be) a matter for the citizens of each country to weigh the greater vulnerability against the advantages.

Let us now consider how the use of science-based technology on a large scale has been linked with an increase in vulnerability.

Modern technology has given Man a power of destruction only possessed higherto by the elemental forces of Nature. Natural catastrophes still occur (the earthquake in China, the Darwin cyclone, etc.) but they are meeting strong competition from the scars left by Man's social and economic development: Malpasset, thalidomide, Seveso and Tenerife are some recent examples, without mentioning the military field where the power of destruction has also grown rapidly.

Actually, two developments are taking place at the same time: natural events are increasingly likely to have devastating effects, and human error can now lead to appalling consequences. These are the result of a phenomenon that accompanied and facilitated economic development, i.e. the concentration of industry in certain cities and regions, and the concentration of population in urban areas near the coasts.* This has greatly increased the dimensions of any problem that arises, so that – as Matthias Haller [23] has noted – most problems now need to be tackled at the national or world level, instead of the family or local level. Concentration is not a direct or a necessary effect of technology, but it corresponds to one of the basic principles of scientific production management and is

* According to Barbara Ward [38], 15 per cent of the world's population lived in towns at the beginning of this century. By 1960 the proportion had reached one-third, and the estimated figure for the year 2000 is 50 per cent.

both a cause and a consequence of higher productivity.

We are now beginning to see that concentration has in some cases gone too far and is producing more ill-effects than benefits. So the advantages of smallness are beginning to attract attention.

In an article entitled "The End of the Cathedrals", the French writer, Jacques Barraux [1] noted that the period of huge factory complexes seemed to be coming to an end. For various reasons (environmental, risk-spreading and labour troubles) an increasing number of manufacturers are giving up large high-productivity plants in order to benefit from the greater operating security of smaller plants. There is even a historical precedent for this — the Great Fire of London in 1666, which revealed the hazards of an excessive concentration of half-timbered housing.

A corollary of concentration is specialization. Here, science-based technology is even more directly involved since, in order to make progress in the knowledge and utilization of physical materials and in economic and social organization, it has been necessary to confine the individual to the specialty in which he has a comparative advantage over others — and this also applies to machinery. Now, a system of highly specialized sub-systems is vulnerable. The sub-systems have no autonomy. If one of them ceases to function, the whole system is paralysed. When the central power station for New York City broke down for twenty-five hours in July 1977, all work in the city had to stop, the criminality rate increased, and the cost of the stoppage was estimated at $5 billion.* Efforts are being made to counter the effect of such failures by making them extremely unlikely, and this aim has been more or less achieved by using still more technology. But this is not the end of the matter. When accidents rarely occur, people are less able to cope with one when it comes.

> "What matters to the user, is the availability of his equipment in working order. Equipment which fails, say, once a year but takes two weeks to diagnose and repair may be far more undesirable for some purposes than equipment which fails once a month but only takes half an hour to repair" ([26], p. 323).

This type of hazard, ranging from disasters to the minor mishaps that may occur to any of us when using modern equipment, is an unwelcome

* *Financial Times*, 15 July 1977.

by-product of technical progress. Not only does the increase of productivity become of doubtful value when the hazard reaches a certain level, but the variance increases and so does public dissatisfaction with technology.

Concentration and specialization are creating problems not only at the macroeconomic level but also for individual firms, where they are increasing the range and extent of entrepreneurial risk.

When a manufacturer decides to set up a factory, he takes a "speculative" risk. If the product is successful on the market, he makes a profit. If it is unsatisfactory, or he miscalculated the demand or the business is badly managed, he sustains a loss and may go bankrupt.

In addition to these voluntarily accepted speculative risks, his business is exposed to a number of "pure" risks that are due simply to the wider uncertainties of existence: his plant may burn down or be sabotaged, his workers may strike, his transport vehicles may be involved in accidents, the computer may break down, his best salesman may die, etc.

For a long time these were routine hazards of business life and required little attention, apart from having the financial side looked after by an employee responsible for insurance policies. The position is quite different today. Risk management has become in the United States, and is becoming in Europe, a major function in the bigger companies [37]. The pure risks have grown to such dimensions in recent years that they are affecting the calculation of entrepreneurial risk. If the installations are destroyed, the cost of replacement will be enormous, quite apart from the loss of business in the meantime, and this is not something that firms can afford to risk in a high-productivity economy. Specialization has reached a level where a single failure in the system (especially the computer [14]) is enough to cause a total stoppage. Workers have been quick to realise this new vulnerability and to take advantage of it in wage disputes. There are also new kinds of pure risk that affect profitability and so increase the speculative risks: the political risk of subsidiaries being nationalized, risk of pollution, and above all the risk of liability for damage or injury caused by the firm's products (in the chemical and pharmaceutical industries, for example). The arrangements for inspection and prevention, handling emergencies and buying insurance cost money, and this is making manufacturers think twice before introducing a new technique, launching a new product, or employing a high-performance process. This kind of cost-benefit analysis is also needed at the national level in order to assess the real returns from technical progress but, as we shall see later, this is not being done and cannot be done in the present state of national accounting.

A reference should be made, before leaving the question of vulnerability, to the views of Rufus E. Miles, who attributes this to the change in the social structure brought about by the new technology and the economic development that followed it. He considers that society has become more vulnerable because the social links within the community have given way, leading to the loss of four behaviour patterns needed for the smooth functioning of the system, namely

> "(1) a high order of compliance on the part of the vast majority of members of key organizations of the society with the decisions of authorized supervisors, managers and officials; (2) a low percentage of acutely alienated persons willing to take major personal risks to sabotage such organizations from within or the larger technostructure from without; (3) a sufficient supply of specialized and qualified manpower, and especially of talented managerial leadership, able and willing to design complex systems and knit together the component operations of the major enterprises; and (4) least appreciated of all, a supportive social structure and body politic on which the separate organizations of the technostructure can depend to furnish the needed reservoirs of manpower, credit, social order, and other requisites of success....A high-energy, high-rise society is immensely more vulnerable to social blackmail than a low-energy one. There are scores, probably hundreds, of specialized occupations that are so essential to the functioning of a large, modern city that their practitioners hold a powerful threat over major segments of the public" ([29], pp. 62–63)

The economic and social system is not, of course, defenceless against the growing risks. The policies to deal with them are of two main types. In the first type, efforts are made to control vulnerability by emphasizing measures for preventing and limiting disasters: electronic monitoring, emergency plans, use of dual systems, strengthening of the enforcement machinery, campaigns to increase safety-mindedness among the public, etc. Such measures can in fact be successful up to a point. Nevertheless, they clearly represent a price that has to be paid for maintaining the system and enabling it to survive since they are designed to limit other costs, i.e. the loss when disasters occur. The other type of policy is directed at the causes of vulnerability. Here the pursuit of productivity is to some extent sacrificed in an effort to find a balance between the advantages and disadvantages of scientific organization of production and society. The abandonment of the search for ever-increasing scale (in the design of factories, towns and transport systems) and the breakaway from moving-belt systems as in the Volvo Works at Kalmar in Sweden fall into this category. This approach admits by implication that the returns to science-based technology are diminishing, and that the optimum is the meeting point where the downward-sloping marginal revenue product curve cuts the upward-sloping marginal cost curve for technical progress. If this optimum is exceeded, the phenomena that we have just described soon appear, and the system itself will generate forces to restore the balance, whether we want it or not. Any policy geared to growth at all costs, along the lines that proved successful in recent decades when the circumstances were quite different, will inevitably be resisted by a growing part of the population, especially the younger age-group who will constitute tomorrow's working population and are therefore an integral part of the structural change taking place.

Current developments throughout the world are significant in this respect. The "technophobia" movement [25], the popularity of the ecological approach to problems, and the frightening consequences foreseen from current basic research -- as in the field of medical biology [27], show that science-based technology has already entered a phase of diminishing returns. The practical effects of this can be seen in the drop in investment in R & D (see Chapter 4), in the hesitations of manufacturers in following up new inventions [8], and in the open opposition to some technical developments (SST, Concorde, nuclear reactors). Hence, the foundation that enabled the Western world to achieve faster and faster growth over the last 200 years has weakened, at least for the moment. Even when national statistics show growth of the GNP, it is only a fictitious growth. The reason why technophobia has recently developed is that welfare had already ceased to increase step by step with economic growth. The weakening of the link between them was to a considerable

extent a reflection of the phenomenon of diminishing returns of technology, but it went unnoticed because of a lack of symmetry in the calculation of the GNP, i.e. the fact that genuine value added and fictitious
value added are treated alike in the national accounts.

5.2. The divorce between economic growth and increased well-being

As a rule, the economic growth of a nation over a given period is more
or less identified with the increase in the total value of all the country's
outputs of final goods and services, plus the balance of factor incomes
from abroad. It differs slightly from the Net National Product (NNP) in
which a deduction is made for depreciation. The NNP is equal to the
national income at market prices, the main elements of which are total
incomes (wages, interest, rents, dividends, profits) and indirect taxes.

As a measure of economic growth, the GNP has for a long time been
criticized by economists on the ground that it takes no account, among
other things, of unpaid work (housework, do-it-yourself, charitable work)
and unreported paid work, or on the ground that the GNP measured at
the national over-all level may conceal regional lags.

More recently, since the publication of books such as the one by Mishan
[30] on the costs of economic growth, it has been pointed out that the
GNP takes no account of external diseconomies caused by industry.*

Obviously an increase in external diseconomies means a diminution of
well-being. If diseconomies are neglected in the calculation of the GNP,
there will be a divergence between economic growth — hitherto regarded
as synonymous with increased well-being — and the actual increase in
well-being. The standard of living measured by the GNP has to be distin-
guished from the level of well-being which is not measured.

In the preceding section we saw that the diminishing returns of technology are reflected in a rapid increase in the external diseconomies from
technical progress. In order to make a proper evaluation of the over-all
effect of technical progress, it is important therefore to find a way of
including these externalities in the accounts.

Broadly, the problem can take three forms: (a) the external diseconomies are associated with a rise in GNP; or (b) they are associated with a
fall in GNP; or (c) they do not affect the calculation of the GNP.

---

* See also Daly [9], pp. 149–152.

The first situation arises when an increase in value added occurs simultaneously with a fall in well-being. It is illustrated in the diagram below, which brings out clearly (if crudely) the distinction between the purely accounting concept of value added, and the combination of psychological, economic and social factors constituting the concept of well-being.

Diagram I

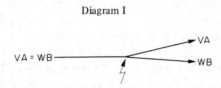

At the starting-point, value added (VA) can be regarded as an appropriate indicator of well-being (WB), but an event or activity producing external diseconomies results simultaneously in a rise in VA and a fall in WB. The drop in the latter is less marked than the increase in VA because of the increased incomes that accompany the latter.

Traffic blocks are a standard example of this situation. They involve loss of time and bad temper, but increase petrol consumption. While this increased consumption increases import demand in the oil-importing countries and so reduces VA, this reduction does not bring VA back to its original level because the additional consumption of petrol leads to an expansion of the country's oil refining and fuel distribution industries.

On the other hand, there is no divergence between VA and WB if the external diseconomies are internalized in the costs of the firm. Diagrams II and III show how this occurs in two stages.

During the first stage, a new product is placed on the market. Other things being equal, the sales of this product naturally cause an increase in value added. However, its production and utilization involve so many nuisances that its effects in terms of value in use and a higher incomes flow are cancelled out, and there is no increase in well-being.

During the second stage, legislation becomes necessary in order to force the producers to adopt anti-pollution safeguards. If these affect the fixed costs of the firms, there is no reduction in final output. The increased cost involves a drop in the profits of the firms responsible for the pollution, but this is offset in the national income by the new incomes arising from the development of an anti-pollution industry. VA is unchanged and WB increases as pollution declines. A further point worth noting is that, by reducing profits, the anti-pollution measures limit the expansion of production that might occur in competitive conditions as a result of new firms entering the field.

Diagram II

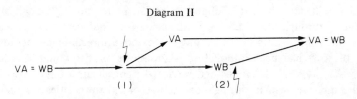

If the impact of the anti-pollution measures falls on the firms' variable costs, the marginal cost of further production increases and the quantity of new products marketed declines. In this case, VA also declines, moving to meet WB which is rising with the disappearance of pollution.

Diagram III

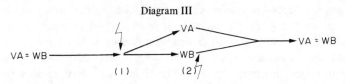

In both cases, production will be abandoned if the cost of the antipollution safeguards is prohibitive. Otherwise, VA and WB both end up at a higher level, indicating that (if we leave aside the problem of measuring utility) the marketing of the new product has increased the total utility.

The commonest situation, however, is where external diseconomies are not reflected in the trend in GNP. Over time, the total of value added in the economy increases but the rising curve of VA diverges from the curve for an ideal indicator of WB (Diagram IV). The divergence may be due, for example, to degradation of the environment.

Diagram IV

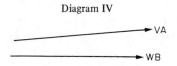

When the divergence is too obvious, there will be pressure from public opinion for the State to reduce the external diseconomies. Usually the cost of this cannot be internalized in the firms as in Diagram III because of current business conditions. The State may therefore finance certain anti-pollution operations, such as the elimination of oil slicks from wrecked tankers. If the additional public expenditure is covered by taxes, there will be a distortion of the allocation of resources: the tax-payers will — through the State — be financing the producers and consumers of the goods and services causing the external diseconomies because — if the cost had been internalized — producers' profits would have fallen or the price of the goods and services to consumers would have risen. The increase in WB through the action of the authorities may in this case be more or less cancelled out by the decline in WB caused by higher taxation. If the additional public expenditure comes from deficit budgeting, resource allocation is still distorted (as the money could have been used for something else) but there is an increase in VA at the same time.

The same is true where households allocate part of their consumption to goods or services that reduce external diseconomies from which they are suffering (e.g. sound insulation near an airport) or repair the damage caused by them (e.g. medical expenses for damage to health).

All such costs are incurred in restoring a level of well-being that has not followed the trend in the national income. They cause a proportionate increase in WB,* but as they increase VA at the same time, the initial gap between the two continues.

Diagram V

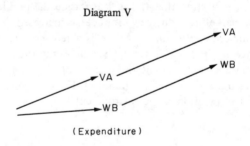

(Expenditure)

Restoration of well-being does not occur here without an increase in VA, as was the case when the costs were internalized. In reality, what is counted

---

* Provided that the expenditure does not in its turn give rise to external diseconomies.

as "value added" is a certain amount of "*value deducted*", indicating the gap between gross value added and actual well-being.

For example, when the authorities in Italy have swimming pools built on the beaches where bathing is banned because the sea is too polluted, it is absurd to count the pools as part of value added. They are built to make good a loss of well-being and not to increase its net value. The fact that some swimmers may prefer the pool to the sea (Fontela [11]) is no justification for including the pool under the heading of "value added". True, for them the combination of polluted-sea plus swimming-pool is a net improvement over the unpolluted sea. But there are other people for whom swimming in a pool is not as enjoyable as sea-bathing, so the two effects cancel out and there is a divergence between value added and well-being. It may be added that a swimming pool built on a beach cannot be put in the same category as a centre for water sports built elsewhere. This example brings out the fact that there are two components in external diseconomies:

(1) A component that is calculable in money terms. This is the amount spent to restore the level of well-being and which is fictitious value added (the cost of the swimming pool).

(2) A component that cannot be so calculated, which represents the loss of amenity after the construction of the pool for those who regard sea-bathing as having a higher utility.

This distinction shows that there is an answer to those who reject any improvement of national accounting systems on the ground that there are no objective criteria for evaluating external diseconomies. It suggests that the national product would more accurately reflect the trend in well-being if the part of the value added produced in overcoming or mitigating external diseconomies were deducted from the gross figure that now appears in the accounts. This would give a net over-all value-added figure as the difference between gross value added and these amounts of value deducted, in the same way as the NNP is obtained by subtracting depreciation. These amounts might well be included under the heading of "depreciation" so as to make the NNP a less imperfect indicator of increase in well-being.

It is quite true that national accountants have never claimed that the GNP measures well-being. But it is also a fact that, in the absence of any other statistical series, the GNP is used in practice to indicate an improvement in the well-being of the population, and that economic policy and numerous pressure groups call for growth in the GNP even if this involves an increase in value deducted.

The question of internalizing external diseconomies in the economic calculus can no longer be evaded. In this respect, we are today in the position prevailing in the nineteenth century before internalization of the cost of industrial accidents (Janice Tait [35]).

5.3. Diminishing returns of technology and economic science

The present increase in "value deducted" reflects the decline in the overall efficiency of technology and, more generally, the difficulty of maintaining the model of economic development inherited from the second industrial revolution. It is becoming increasingly obvious that economic activity is an integral part of the environment, and that the mechanistic picture of the circular flow − as reproduced in all economic textbooks − is misleading.

This fact can perfectly well be recognized without shedding tears over the "appalling degradation" inflicted on nature by "evil" technocrats, producers and consumers. It does not imply a conflict between economics and ecology. What is needed is an awareness that the interaction between the economic and the ecological spheres may at some time in the future involve concrete and calculable losses in the economic sphere, such as a decline in productivity or even in output (as in the fishing industry) and not only immaterial costs in the form of nuisances.

The diminishing returns of technology confirm the essentially biological nature of the economic process. Alfred Marshall was the first to realize that the subject matter of economics is not a mechanical system, but an evolving reality in which adaptation to novelty plays an essential part. According to Nicholas Georgescu-Roegen [15], the "arithmomorphic" conception of economics can mainly be traced back to Jevons and his "mechanics of utility and self-interest" and to Pareto's notion of *homo oeconomicus*. It has been responsible for the constant preoccupation with tracing a demarcation line or "iron curtain" between what belongs to economics and what lies outside. As a result, the phenomenon of technology is not directly studied, only being introduced into the economic discussion as a datum and not as a variable.But real life cannot be reduced to a series of discrete, real number concepts. The economic field is linked with the political, social, psychological, cultural and other fields in a dialectic process. Hence, economics confines itself more and more to the short term, the period when as many variables as possible are data, and mechanical and quantitative analysis can be meaningful. Of course, this approach is useful and even essential — up to a point. It enables theories to be constructed and tested. When there are no stable phenomena there can be no theory. But it also makes it possible for the underlying phenomena responsible for the long-term trend and for the malfunctioning of short-term models, e.g. the decrease in the productivity of technology, to be ignored.

To sum up, technology shows the signs of what are probably two processes of a biological, evolutionary nature:

(1) A process of ageing with gradual loss of power and impetus (Chapter 4). This fits in with Landes' analysis (Chapter 2) of the cyclical nature of economic development and Gimpel's thesis which on many points is reminiscent of Oswald Spengler's. Gimpel writes: "A misreading of the history of technology led our present-day society to think that we were witnessing a continuous development of science and technology. It is the duty of the historian to correct this belief. Our Western civilization is at the moment on a technological plateau that will last well into the third millenium" ([19], p. 229).

(2) The survival difficulties of a system that grew up rapidly and is now suffering from its very size and success, and provoking rejection symptoms in the natural and human environment on which its growth depends (Chapter 5). This fits in with recent thinking on the economic process.

N. Georgescu-Roegen points out that the economic process cannot be compared to a mechanical system but resembles a thermodynamic system exchanging energy with the outside (i.e. the extra-terrestrial universe) in the form of solar energy but not matter.* Like any thermodynamic system that has not reached an equilibrium where temperature is uniform throughout the system, the economic process produces entropy — indeed

* i.e. to use Kenneth Boulding's expression [5], the economy of "spaceship Earth".

it is accelerating the planet's natural entropy — by continuously trans-forming* the usable energy in organized matter into unusable energy in the form of waste.** This is an application to the economy of the second law of thermodynamics (or the law of entropy) whereas economics had retained only the first law, i.e. the principle of the conservation of matter and energy.

His line of thought can be illustrated by the following quotations:

> "The economic process consists of a continuous transforma-tion of low entropy into high entropy, that is, irrevocable waste or, with a topical term, into pollution" ([15], p. 281).

> "The truth is that every living organism strives only to main-tain its own entropy constant. To the extent to which it achieves this, it does so by sucking low entropy from the environment to compensate for the increase in entropy to which, like every material structure, the organism is con-tinuously subject. But the entropy of the entire system — consisting of the organism and its environment — must in-crease" ([16], p. 41).

> "The entropy of copper metal is lower than the entropy of the ore from which it was refined, but this does not mean that man's economic activity eludes the entropy law. The refining of the ore causes a more than compensating increase in the entropy of the surroundings. Economists are fond of saying that we cannot get something for nothing. The entropy law teaches us that the rule of biological life and, in man's case, of its economic continuation is far harsher. In entropy terms, the cost of any biological or economic enterprise is always greater than the product. In entropy terms, any such activity necessarily results in a deficit" ([16], pp. 41–42).

This also explains why recycling operations are, in purely physical terms, bound to fail.

The entropic nature of the economic process is, however, different from a purely physical relation in that the true output of the economic

---

* The word "entropy" is derived from a Greek word meaning "transformation".
** See earlier footnote in connection with pollution (p. 91).

process is not entropy but an immaterial flux: the enjoyment of life.

This analysis leads obviously to an extremely pessimistic view of humanity's future (cf [17], p. 35), but there is consolation in the thought that entropy is an extremely long-term phenomenon on the planet's life scale, and that the acceleration of entropy due to economic development will not perhaps fundamentally change the position.

Moreover, the latest developments in thermodynamic theory as regards irreversible phenomena suggest the possibility of biological states in which the production of entropy is minimal. On the basis of relationships worked out by Onsager, Prigogine's theorem [32] shows that, subject to certain conditions, the production of entropy is at its minimum in a stationary state. These developments naturally strengthen the convictions of writers such as Daly [9], who follow John Stuart Mill in arguing in favour of the stationary state. However, the conditions in which Prigogine's theory are valid are rather strict:

> "The scope of Prigogine's theory of minimum production of entropy is very general since it applies to all non-equilibrium stationary states, whatever the nature of the constraints. It must, however, be emphasized that the theorem is also subject to severe restrictions since it only relates to the area of linear thermodynamics of irreversible phenomena, or even more narrowly to the region where phenomenological coefficients are assimilable to constants and satisfy Onsager's relations (Glansdorff and Prigogine [20], p. 47).

The outlook in the medium-term is less dark. Of the two processes that are characteristic of the diminishing returns of technology, the first in particular could only be transitory and a new cycle of technology could then have a beneficial influence on the second process. The views presented here are not intended to prove that science and technology are condemned for ever to follow the law of diminishing returns. Our aim is to gain an overall view of a large number of developments over the past decade, which are sufficiently important to merit our attention but which need to be verified and confirmed by other studies of a less general character.

It is true that the experience of technological development in the nineteenth century showed that inventions are a stimulus to economic growth only where intellectual and economic conditions are favourable. In this respect, present conditions are — to put it mildly — not particularly favourable. In the industrialized countries, saturation point has to all intents and purposes been reached in the field of consumer durables. In the developing countries, effective demand is insufficient. And everywhere dissatisfaction with the techno-industrial system is openly expressed.

Nevertheless, this situation may change. A new take-off may be stimulated by new inventions, in line with Schumpeter's theory of technological cycles [33]. Perhaps the innovations will occur in the domain of services, in which case they would counterbalance the negative effects of the tertiarization of the economy discussed earlier. The main hope lies in possible progress in the solar energy field and in methods of storing electrical energy. Development of solar energy would indeed avoid the main external diseconomies from present-day science and technology by offering a non-polluting source of energy available in the form of a scattered flow and not of a concentrated stock. This feature would enable us, among other things, to move from a centralized and specialized model of society to a more decentralized and autonomous model, so as to overcome the tendency to increasing vulnerability, even at the cost of an *apparent* loss of productivity.

It is obvious that such a change will take time because of the delay factor operating in the field of modern R & D (cf. Chapter 4). Meanwhile, we must be fully conscious of the limits to which we are subject. Recent economic growth can be traced back to the historic marriage of science and technology in the nineteenth century. This had a very specific background, and the process is not therefore one that can easily be reproduced. The marriage will probably figure in history as one of the greatest achievements, for it is bringing us — as we now know — to a totally new era.

In their efforts to suggest remedies for the increasing vulnerability, those concerned with long-term political issues should bear these realities constantly in mind.

References

1. Jacques Barraux. "La fin des cathédrales", *Economia,* February 1977, pp. 55–57.
2. Petr Beckmann. *The Health Hazards of Not Going Nuclear.* The Golem Press, Boulder, Colorado, 1976.

3. Daniel Bell. *The Coming of Post-Industrial Society.* Basic Books, Inc., New York, 1973.

4. Antoinette Boegner. "Les services dans l'économie: la fin de l'anarchie, le début de la sagesse", *Les Dossiers d'Entreprise,* February 1973, pp. 30–41.

5. Kenneth Boulding. "The economics of the coming spaceship Earth", in [9], pp. 121–132.

6. Peter Chapman. "Industry and the environment", *New Scientist,* 5 May 1977, pp. 256–258.

7. Colin Clark. *Les conditions du progrès économique.* P.U.F., Paris, 1960.

8. François Dalle. "L'entreprise face à l'imprévisible". Lecture given at the International Centre for Monetary and Banking Studies, Geneva, 15 December 1976.

9. Herman Daly. *Toward a Steady-State Economy.* Freeman & Co., San Francisco, 1973.

10. Jean Denizet. "Chronique d'une décennie", in Perroux, Denizet et Bourguinat, *Inflation, Dollar, Eurodollar.* Gallimard, Paris, 1971, pp. 23–104.

11. Emilio Fontela. "Energie et économie nationale", Study Group on Energy, University of Geneva, 28 April 1977.

12. Jean Fourastié. *Le grand espoir du XXe siècle.* Gallimard, Paris, 1963.

13. Octave Gelinier. "A la découverte du nouveau tertiaire", *Le Monde de l'Economie,* 20 April 1976.

14. André George. "Nature et importance des pertes économiques dans l'utilisation de l'informatique en Europe en 1985", *Geneva Papers on Risk and Insurance,* No. 3, October 1976.

15. Nicholas Georgescu-Roegen. *The Entropy Law and the Economic Process.* Harvard University Press, 1971.

16. Nicholas Georgescu-Roegen. "The entropy law and the economic problem", in [9], pp. 37–49.

17. Nicholas Georgescu-Roegen. *Energy and Economic Myths.* Pergamon Press, New York, 1976.

18. Giarini, Loubergé, Schwamm. *L'Europe et les ressources de la mer. Introduction à l'economie marine.* Ed. Georgi, St. Saphorin, Switzerland, 1977.

19. Jean Gimpel. *La révolution industrielle au moyen-age.* Seuil, Paris, 1975.

20. Glansdorff et Prigogine. *Structure, stabilité et fluctuations.* Masson, Paris, 1971.
21. Mark Greene. *Risk, Insurance and the Future.* Indiana University Press, 1971.
22. Philippe Grosjean. *La Tertiarisation de l'economie française.* IUEE, Geneva, June 1977.
23. Matthias Haller. *Sicherheit durch Versicherung?* H. Lang, Berne, 1976.
24. Garrett Hardin. "The tragedy of the commons", in [9], pp. 133–148.
25. Hal Hellman. *Technophobia – Getting out of the Technology Trap.* Evans & Co., New York, 1976.
26. Stanley Hurst. "Beware of the general purpose micro-processor", *New Scientist,* 10 February 1977, pp. 322–324.
27. Leon Kass. "The new biology; what price relieving man's estate?", in [9], pp. 90–113.
28. James Meade. *The Theory of Economic Externalities.* Sijdthoff, Leiden, 1973.
29. Rufus Miles. *Awakening from the American Dream. The Social and Political Limits to Growth.* Universe Books, New York, 1976.
30. Ezra Mishan. *The Costs of Economic Growth.* Pelican Books, London, 1967.
31. Jacques Parent. "Les causes structurelles de l'inflation", paper for the Congrès International des Economistes de Langue Française, Paris, May 1977.
32. I. Prigogine. *Thermodynamics of Irreversible Processes.* Wiley, New York, 1955.
33. Joseph Schumpeter. *Business Cycles – A Theoretical, Historical and Statistical Analysis of the Capitalist Process.* McGraw Hill, New York, 1939.
34. Robert Solow. "The economics of resources or the resources of economics", *American Economic Review,* May 1974, pp. 1–14.
35. Janice Tait. "Non-renewable resources – what alternatives?", *Canadian Institute for Public Affairs,* Environment, Canada, Planning and Finance Occasional Paper No. 3, 1975.
36. Alvin Toffler. *Future Shock.* Random House, New York, 1970.
37. Varii auctores. "The management of risk and insurance", *Geneva Papers on Risk and Insurance,* No. 2, August 1976.
38. Barbara Ward. *The Home of Man.* Norton, New York, 1976.

CHAPTER 6

Some Concluding Reflexions on Historical and Political Aspects

The preceding discussion has enabled us to gain some idea of the phenomenon of diminishing returns of technology, and of technology's influence
on the origin and subsequent history of economic growth over the last
centuries.

In the course of that period, countries in Europe and elsewhere have
made the change from a pre-industrial to a post-industrial structure. We
are now seeing a globalization of the industrialization process. While this
is not progressing evenly enough to satisfy the aspirations for a more
equitable distribution of the world's wealth, it has certain analogies with
the spread of the Industrial Revolution to continental Europe in the last
century. The similarity is obvious on the political level. As in the earlier
case, there is an irresistible upsurge of independence movements in the
regions touched by industrialization. This corresponds to a spreading of
the European ideological model born of the Industrial Revolution (in
its liberal and Marxist versions). One hopes, however, that in the Third
World the model will take a form that avoids a repetition at each stage
of the set-backs that occurred in Europe. One may perhaps be forgiven for
thinking, moreover, that world equilibrium is more likely to be achieved
through the development of the different national cultures rather than
through a standardization of culture.

In the case of the European countries, their determination to maintain
the individuality of Europe is forcing them to tackle the first stage in
what could become a pluralist and worldwide form of society. As Denis
de Rougemont has said, every activity in the world of culture is a form
of production. Europe must therefore re-acquire the ability to "produce"

by seizing the opportunity of transforming the present economic crisis into a "challenge", in the sense in which this term is used in Toynbee's study of civilizations.

What Europe has been experiencing since 1968 is above all a crisis due to the ending of an epoch. The crisis has been marked by three major events:

1. The intuitive realization, at the time of the spontaneous movements in 1968, that growth in GNP was no longer bringing a proportional increase in well-being.
2. A sudden awareness, inspired by the Club of Rome, of the *external* limits to growth, which the dominant ideology of the preceding thirty years or so had caused to be forgotten.
3. A realization that technology is subject to diminishing returns and that these are the *internal* limits to growth.

The marriage between science and technology has operated in two directions. On one hand, from the end of the last century it provided the motive force for a period of economic growth in which the rates achieved after the Second World War had no parallel in human history for a period of such length. On the other hand, the marriage also explains the structural origin of the present depression. Phenomena analogous to those involved in the law of diminishing returns to the factors of production are causing the fertility of the marriage to decline, through an evolutionary process over time that is reminiscent of biological processes. Hence, economic growth can, in the medium and long term, only return to the rate for the whole period of the industrial revolution, i.e. about 2 per cent per annum. Even this is an optimistic assumption, particularly if one is thinking of this rate in terms of actual growth in material welfare measured in net value added.

In view of the foregoing, any short-term economic policy based on an assumed need to provoke an over-all revival of demand inevitably leads in a short time to inflation and new austerity measures. One need have no hesitation in saying that the advocates of a policy of deficit state financing and easy money are one world war behind in their thinking. The circumstances in which Keynesian policies were successful over a period of thirty years are not those of today (apart from the impropriety of limiting Keynes' theory to regulation of demand alone). At the time of the Great Depression and in the aftermath of the Second World War, technology offered vast scope for expansion of supply. Apart from a few special cases, this is no longer generally true. It is useless to stimulate demand unless there is reason to believe that supply will follow it.

Some writers, especially in the business press, have begun to mention this point. A recent study by *Business Week*** shows that, since 1960, the recovery of investment after recessions has been slower each time. On the whole, however, the commentators have drawn from the study what we believe to be mistaken conclusions. The slackness of investment is often attributed to lack of enterprise on the part of industry today. But how can an executive in the chemical industry be expected to react, after twenty or thirty years of boom in synthetic fibres, when he knows that there will probably be nothing in the coming years to compare with the introduction of nylon, acrylics and polyesters?

Other advocates of a reflation policy point out that firms have unused capacity, and there are reports from time to time that industry has been operating since 1973 at 80 per cent of capacity. But average capacity utilization rates are meaningless. The argument overlooks the fact that the fabrication of a product requires several stages. Once the production capacity at one of the stages is fully employed, this prevents final output from being increased any further, even if there is underutilization of total capacity.

Now, elimination of such "bottlenecks" may not be an easy matter. Some are due to hard, economic facts and not to the alleged inability of managements to respond to excess demand. They may occur at the import stage for raw materials, in the production of intermediate goods or in the distribution network; and may persist because technology has not provided any new solution, or because the expansion involved would create on balance more problems that it would solve, especially if labour and ecological factors are taken into account.

Hence, while awaiting a hypothetical new technological cycle, the monetarists' rule of moderate, regular growth of the money supply seems the most sensible approach in present circumstances, particularly as the slow-down in growth and the increase in "value deducted" both cause budget deficits to swell automatically.

---

* "Capital spending is making its slowest post-war come-back from recession", *Business Week,* 14 September 1976, p. 64.

A new structural cycle can only occur — and with ever-increasing time-lags — after a new breakthrough in fundamental research (e.g. one that made it possible to produce, store and distribute energy at very low cost, so increasing operating autonomy within the system). This implies that, in industry, the productivity of technology will have to be measured in terms of improved autonomy and survival capacity (reduced vulnerability) rather than in terms of speed of production, economies of scale and further concentration. The original meaning of "welfare" will have to be restored, and based on the actual value in use of the planet's human and material resources.

A new and creative Europe can emerge from this epoch. The post-industrial society is beginning to take shape, though far-removed from the picture proposed by those, like Daniel Bell, who launched the concept over twenty years ago.

Let us consider some of the pointers:

1. From the structural angle, the Industrial Revolution transformed a predominantly agricultural society into a society in which the emphasis was increasingly on industrialization and the new pattern of production. In the new industrial phase, it is the tertiary sector (commercial and personal services) which is developing, particularly *within* the secondary and primary sectors. A "tertiarization" of the economy is taking place, not only through the expansion of the tertiary sector itself but also through a tendency for the other two sectors to become sectors with tertiary characteristics.

2. As a result, physical production has become secondary to sales, distribution and related services. Whether one considers meat, textiles or any other product, the cost of actual production is often less than 20 per cent of the price to the final consumer. Distribution, warehousing, management, etc., make up the bulk of the costs or, to put it another way, the bulk of the energy needed for the functioning of the system.

3. In the industrial development of societies technology has been a decisive factor. Since its main impact has been in the area of physical production, one might say that it only affects part of the economic structure. This is true, but there is another side to the picture, namely, that technology has caused industry to develop in a way in which "tertiarization" was inevitable. The achievement of higher

productivity in terms of physical output has involved direct and indirect costs that often equal or exceed the initial saving from productivity.

4. The tertiary sector is also changing. Service industries are making increasing use of "hardware" (vehicles, office machines, computers) and are thus becoming more like the manufacturing sector with the worker serving the machine. This is why the pictures of post-industrial society suggested by Daniel Bell, Jean Fourastié and others are out of date. What is happening now is not something that can be described by mechanically extrapolating from a classification of the economy into three sectors and assigning a period of historical predominance to each. The tertiarization of the physical production sectors is a sign that the old system dating back to the Industrial Revolution is breaking up, and that we are moving into a new economic system in which all the basic concepts and parameters (value, production function, capital, assets, production structure, demand, etc.) are changing. Hence, a great deal of empirical research is needed on the development and spread of technology.

5. Science-based technology has contributed not only to the increase in value added (as a measure of GNP); it has also gradually stimulated the growth of "value deducted". Pollution and the annoyances of industrial civilization make the economic machine run even faster to restore the balance upset by the system itself (as in the case of swimming pools built on the beaches of the polluted sea). The gap between rising GNP and increased well-being is becoming wider and calls for a reconsideration of the economic concept of "value".

6. Developments in science and technology are no longer linked together in the same way as they were during the twenty years following the Second World War. We are probably in a declining phase of a long cycle of several decades, marked by a slow-down in technical progress. We cannot expect breakthroughs in the near future comparable with those in aviation, plastics and synthetic fibres.

All of which goes to show that the law of diminishing returns applies also to technology.

What conclusions can be drawn as regards the political and socio-cultural fields?

The doctrines of liberalism and socialism have dominated the period of industrial maturation in Europe. There has been and still is what might be described as a civil war between them in the field of ideas. Their emergence was closely bound up with the development of a predominantly manufacturing economy, and in a sense it can be said that economies of scale and concentration of production created the material basis for the concentration of power.

The two related ideologies were the result of a compromise — in a given historical and economic context — between the democratic aspirations of mankind and the harsh realities of its journey towards higher forms of civilization. With the gradual disappearance of the historical and economic context from which they emerged, the power systems claiming legitimacy from one or other of the ideologies are being pushed in two opposite directions by

(a) A trend towards a reconciliation between the two ideologies.

(b) A trend towards open opposition to authority where its ideological justification has become unreal and its continuance is based solely on *realpolitik*. The class struggle, moving in a direction that began with Lenin, has also changed into a struggle between élites in the name of different groups but without adequate participation of the people.

It seems clear that any political scheme based on the logic of concentration of power — however justified it may be in other respects — will inevitably lead to a struggle between ruling classes and so to paternalism, authoritarianism or, worse still, dictatorship. Hitler and Stalin were not simply tragic accidents; they were the culmination of an *identical* process, operating in the favourable environment of a concentrated socio-economic system and an uprooting of the population that left the individual at the mercy of unrestrained demagogy. What is the likely course of events?

Our discussion of the diminishing returns of technology shows that the main emphasis is no longer on economies of scale and concentration. On the contrary, we can see the beginnings of a move toward a pattern of increasingly autonomous units.

From the political point of view, therefore, the current trend in the economy is paving the way for a development of federalism, i.e. a pluralist organization of autonomous member units. The framework for a possible "historic compromise" between the changing pattern of industry and the demand for greater liberty is now developing at this federal level. In these circumstances, the battle for a united Europe is not and cannot be a battle for abstract institutions incapable of resisting the only two realistic alternatives, i.e. political bargaining at the nation-state level, or a Napoleonic vision of Europe where the nation-state is replaced by a supra-national authority. European democracy can be achieved only in opposition to the nation-state and to any such super-state.

Federalism thus offers us a chance of laying the foundations for political democracy in the new post-industrial age. We should be searching in the present for the germs of future developments, rather than keeping our eyes fixed on what will soon become part of past history.

Index

119